Research at NaMLab

Band 2

Research at NaMLab

Band 2

Herausgeber:

Prof. Dr.-Ing. Thomas Mikolajick

Guntrade Roll

Leakage Current and Defect Characterization of Short Channel MOSFETs

Logos Verlag Berlin

Research at NaMLab

Herausgegeben von
NaMLab gGmbH
Nöthnitzer Str. 64
D-01187 Dresden

Bibliografische Information der Deutschen Nationalbibliothek

Die Deutsche Nationalbibliothek verzeichnet diese Publikation in der Deutschen Nationalbibliografie; detaillierte bibliografische Daten sind im Internet über http://dnb.d-nb.de abrufbar.

ISBN 978-3-8325-3261-1
ISSN 2191-7167

Logos Verlag Berlin GmbH
Comeniushof, Gubener Str. 47,
10243 Berlin

Tel.: +49 (0)30 / 42 85 10 90
Fax: +49 (0)30 / 42 85 10 92
http://www.logos-verlag.de

Leakage Current and Defect Characterization of Short Channel MOSFETs

Charakterisierung von Leckströmen und Defekten in Kurzkanal-MOSFETs

Der Technische Fakultät der Universität Erlangen-Nürnberg zur Erlangung des Grades

DOKTOR-INGENIEUR

vorgelegt von
Guntrade Roll
Erlangen, 2012

Als Disstation genehmigt von
der Technischen Fakultät der
Universität Erlangen-Nürnberg

Tag der Einreichung:	20.03.2012
Tag der Promotion:	10.09.2012
Dekanin:	Prof. Dr.-Ing. Marion Merklein
Berichterstatter:	Prof. Dr. rer. nat. Lothar Frey
	Prof. Dr. Ing. Thomas Mikolajick

Prints of the PhD can be ordered:

Leakage Current and Defect Characterization of Short Channel MOSFETs
Logos Verlag Berlin:
http://www.logos-verlag.de/cgi-bin/engtransid?page=Buchrcihen/namlab.html&lng=deu&id

Acknowledgment:

This work was successful because of the people mentioned here, and many more. Due to the Qimonda insolvency, I had some turbulent times in the middle of my thesis. I am very thankful for the all the support during this phase.

First of all, I want to name the person who guided me from the beginning to the end: Matthias Goldbach. Together with him, I want to thank all the colleagues from Qimonda who not only provided me with samples but also with many fruitful discussions. Sven Finsterbusch and Adriana Sanchez who helped me to understand the measurement structures, and C. D. Nguyen who started me up with the simulation. At Namlab, I want to especially thank Stefan Jakschik and Andre Wachowiak who took the time to read through all my publications including this thesis. But all my co-workers at Namlab somehow influenced this study by improving the measurement setup, or debating the results of the experiments. The time I got to work on my PhD thesis with the Namlabsen was one of the best I had. The simulation model that concludes my thesis is inspired by Alexander Burenkov, and the simulation group of the Fraunhofer IISB in Erlangen. I also want to thank A. Leuteritz, K. Mothes, S. Jansen, M. Mildner for TEM analysis, and H. Hortenbach, and M. Ogiewa for SIMS measurements.

The persons who made it possible that my PhD was continued without interruption are: My supervisor Prof. Lothar Frey, Prof. Thomas Mikolajick, Dr. Alexander Ruf, and Dr. Jürgen Rüstig. At last, I want to thank my parents and friends who kept my feet on the ground, and corrected my spelling.

Abstract

The continuous improvement in semiconductor technology requires field effect transistor (FET) scaling while maintaining acceptable leakage currents. High-k dielectrics, and metal gates are used to reach a low equivalent oxide thickness of the gate stack, and a low gate leakage. Many process changes are needed to integrate hafnium silicon oxide as new gate dielectric in peripheral transistors of state of the art dynamic random access memory. This study investigates the influence of important process changes on the intrinsic transistor leakage currents. The defects responsible for trap assisted leakages are analyzed by electrical measurements, and simulations. P-channel transistors with hafnium based gate oxide, and titanium nitride electrode have a reduced doping concentration of the bulk, and an aluminum oxide layer in the gate stack to reach reasonable threshold voltages. Reduced transistor bulk doping decreases gate induced drain leakage (GIDL) accompanied by an increase in subthreshold current. The subthreshold current increase is partly compensated in devices with reduced equivalent oxide thickness due to a better control of the channel. The transistors with metal gate, and aluminum oxide layer have a higher silicon to gate dielectric interface defect density. A change in source/drain extension spacer material from silicon oxide to silicon nitride is needed to avoid unintended bird beaks. This change leads to a rise in electric field at the gate edge, and higher GIDL currents. Optimum results are achieved with a two layer silicon oxide and silicon nitride source/drain extension spacer. Germanium implantation is needed to preamorphize hafnium silicon oxide in order to remove it after gate patterning. A low implant energy is beneficial because the damage is close to the silicon surface, and can be healed efficiently by annealing. In addition to the introduction of hafnium silicon oxide in the gate stack, scaling also requires a reduction in lateral source/drain dimension. A controlled overlap between source/drain and gate is also required. Carbon is chosen to reduce the boron transient enhanced diffusion of the p-channel FETs. An increase in carbon implantation dose of 1.15 leads to a factor of 1.3 more electrical active defects in the bulk silicon around the source/drain. The density of silicon to gate dielectric interface defects at the gate edge is also increased. The additional defects lead to a higher GIDL, and source/drain extension leakage. Improvement of etching conditions, and thermal treatment significantly reduce the diffusion of interstitial defects in the source/drain extension depletion region.

Zusammenfassung

Für einen stetigen Fortschritt in der Halbleitertechnologie werden skalierte Feldeffekttransistoren mit niedrigen Leckstömen benötigt. Dielektrika mit hoher Permittivität (high-k) und Metall Elektroden werden genutzt, um eine geringe äquivalente Oxiddicke ohne erhöhte Gateleckströme zu erreichen. Viele Prozessänderungen sind notwendig um Hafniumsiliziumoxid als Dielektrikum in die Peripherietransistoren von DRAM Speichern einzufügen. Diese Arbeit untersucht die Einflüsse der verschiedenen Prozessänderungen auf die intrinsischen Leckstöme der Transistoren. Defekte, welche die Leckströme erhöhen, werden mit elektrischen Messungen und Simulationen charakterisiert. Um die Einsatzspannung in high-k p-Kanal Transistoren mit Titannitrid Elektrode einzustellen, ist eine Verringerung der Dotierung im Substrat und eine Aluminiumoxid Zwischenschicht im Gate notwendig. Die verringerte Dotierung im Substrat führt zu einem niedrigeren Gate induzierten Drainstrom (GIDL) und höheren Unterschwellstrom. Die Erhöhung des Unterschwellstromes kann in Transistoren mit niedriger äquivalenter Oxiddicke teilweise durch die verbesserte Kontrolle über den Kanal kompensiert werden. Die Transistoren mit Metallelektroden und Aluminiumoxid Zwischenschickt zeigen einen Anstieg der Silizium zu Gate Dielektrikum Grenzflächendefekten. Die Umstellung von Siliziumoxid zu Siliziumnitrid als Material für den Source/Drain Extension Abstandshalter verhindert die lokale Oxidation des Dielektrikums an der Kante des Gates. Dies führt aber zu höheren elektrischen Feldern an der Gatekante und somit zu steigenden GIDL Strömen. Ein zweischichtiger Abstandshalter aus Siliziumoxid und Siliziumnitrid senkt den GIDL Stromverbrauch, ohne dass eine Aufdickung des Dielektrikums zu beobachten ist. Ein Amorphisierungschritt mittels Germanium Implantation ist notwendig, um das Hafniumsiliziumoxid während der Transistorstrukturierung wieder von den Source/Drain Regionen zu entfernen. Eine geringe Implantationsenergie ist vorteilhaft, da dann die meisten Defekte nahe an der Siliziumoberfläche entstehen und thermisch gut ausgeheilt werden können. Neben der Umstellung auf Hafniumsiliziumoxid als Dielektrikum ist es auch wichtig die laterale Ausdehnung der Source- und Drainregion zu reduzieren und die Überlapplänge der Source/Drain Region mit dem Gate zu kontrollieren. Kohlenstoff wird deshalb implantiert, um die transient erhöhte Diffusion des Bors in den p-Kanal Transistoren zu unterbinden. Eine Erhöhung der Kohlenstoffdosis um den Faktor 1.15 führt zu einem Anstieg der elektrisch aktiven Defekte in der Source/Drain Extensionverarmungszone um den Faktor 1.3. Zusätzlich wird auch die Anzahl der Defekte an der Dielektrikumsgrenzfläche nahe der Gatekante erhöht. Diese Defekte führen zu einem Anstieg im GIDL und Source/Drain Extensionleckstrom. Eine gezielte Verbesserung der Gateätzprozeduren und der anschliessenden thermischen Prozessschritte kann die Diffusion von Defekten in die Source/Drain Extensionverarmungszone verringern.

Contents

1 Introduction

Aggressive scaling reduces costs, increases speed, and enables a higher density of microelectronic metal oxide semiconductor field effect transistors (MOSFET) on a chip. Over the last decades, the amount of transistors per chip has roughly doubled every 18 months by shrinking both horizontal and vertical dimensions (Moores Law) [1, 2]. For MOSFET shrinking, gate length, gate width, oxide thickness, and junction depth have to be reduce [3].

Different scaling rules have been suggested. Constant field scaling preserves the electrical field in the device while shrinking [4]. This is achieved by reduction of the supply voltage, and increase in doping concentration [3]. Constant field scaling leads to reliable, fast, and small devices while the power density is constant. One problem in constant field scaling is that the transistor subthreshold swing is not improved. If the power supply voltage of the device is decreased as proposed, it approaches the threshold voltage. The on-current reduces, and the off-current stays constant. This makes it harder to distinguish the transistors on-, and off-state. The constant voltage scaling, on the other hand, leads to an increase in power dissipation, and reliability problems [5]. Generalized scaling rules are introduced, where the vertical, and lateral electric field increase by the same factor [6]. In this case the power supply voltage is scaled moderately.

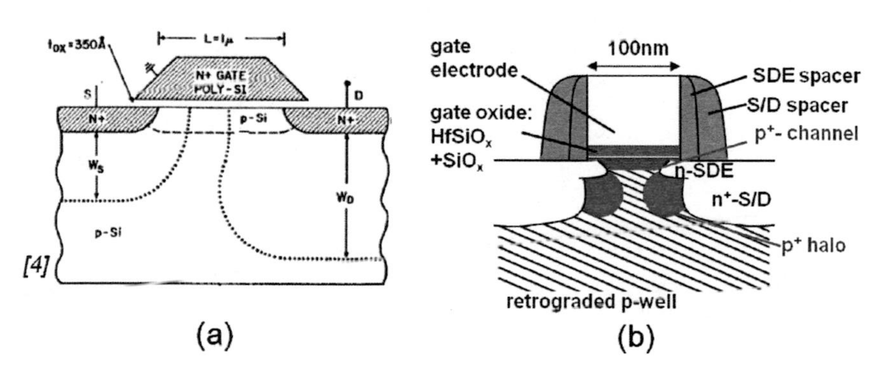

Figure 1.1: Basic sketch of a:
(a) $1\mu m$ long NFET from 1974 [4].
(b) 100nm long NFET investigated in this study.

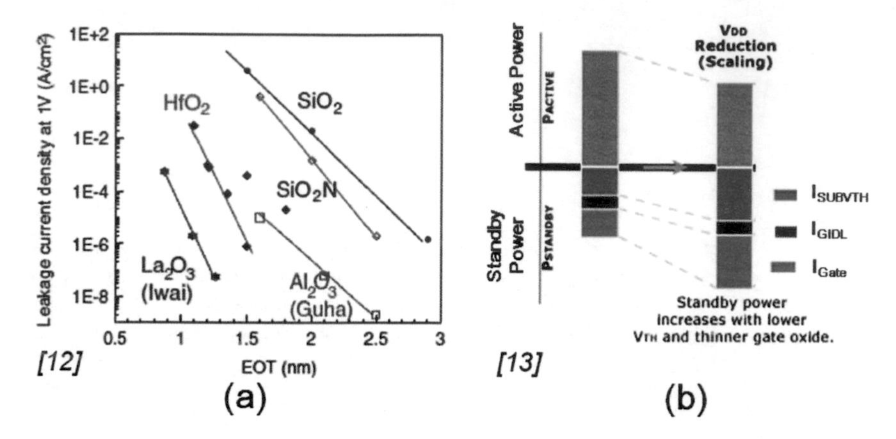

Figure 1.2: (a) Gate leakage for different dielectric materials as a function of the electrical equivalent oxide thickness (EOT) [12].
(b) Power consumption change by scaling from 130nm to 90nm technology node for high performance devices is shown. Standby power is increased by subthreshold leakage (I_{SubVth}), gate induced drain leakage (I_{GIDL}), and gate leakage (I_{Gate}) through a silicon oxynitride [13].

Figure 1.1 shows the basic sketch of a 1μm long transistor of 1974 [4], and a 100nm long transistor used in this study. Many process changes have been implemented to meet the challenges of scaling. For example, shallow source/drain extensions (SDE) are implanted to ensure a good connection to the channel, and to reduce short channel effects. In this way, the depth and doping of the source/drain (S/D) can be increased to reduce the spreading- and contact resistances of the devices which limit the on-current [7]. The introduction of halo implants in the transistor bulk also reduces short channel effects [8]. The gate tunneling current is minimized by an insulator material with a high permittivity, called high-k dielectric [9–11] (Fig. 1.2(a)). Polysilicon gate depletion is avoided by implementing metal electrodes in the gate stack [12].

Leakage currents are one main problem of scaling (Fig. 1.2(b)). Also reliability, and mobility reduction by higher electric fields are detrimental side effects [3]. There are some publications analyzing transistor leakages [14–16]. Trap assisted leakage current mechanisms highly enhance power consumption.

Effects of key innovations, such as the introduction of high-k metal gates, and shallow junctions, on the intrinsic leakage currents of scaled transistors are studied during this investigation. The leakage current paths are separated, and the leakage mechanisms are analyzed. The defect distribution in the MOSFETs is estimated comparing electrical measurements, and simulations. Additional physical characterization of the devices is made to confirm the conclusions extracted from the electrical measurements.

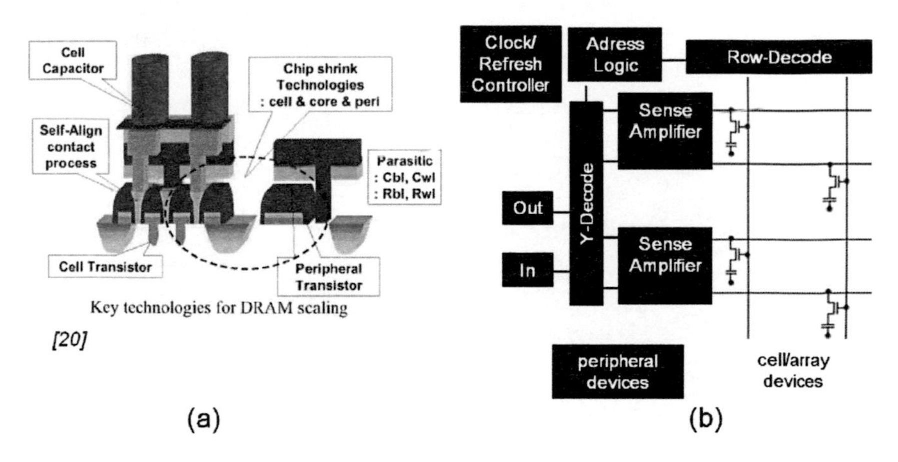

Figure 1.3: (a) Key elements and devices influencing DRAM downscaling are displayed as schematic [20].
(b) The sketch shows the peripheral DRAM circuitry and the array devices [21].

DRAM peripheral MOSFETs on fully processed wafers, of the buried word line technology of Qimonda [17], are investigated. The DRAM is a volatile memory that reacts very sensitive to leakages [18]. High off-currents reduce the refresh time. Low standby power DRAMs are beneficial for mobile, and embedded electronics. The DRAM periphery consists of the sense amplifier that boosts the signal coming from the memory cell, and the logic responsible for addressing each memory cell, input, output of data, and internal refresh of the DRAM (Fig. 1.3). The transistors analyzed in this study are from the logic peripheral devices. The word line driver requires MOSFETs with a thick gate dielectric because the transistors are used at high voltages (3V). For these high voltage devices, leakage and reliability are especially critical. Fast short channel transistors with high on-currents at supply voltage are used for logic circuits. They are called regular peripheral MOSFETs. Planar structures are commonly used today for peripheral DRAM transistors [19, 20].

The DRAM transistors can not be optimized for the scaling alone. They have to withstand the complete DRAM process flow [22]. Among others, this includes a long time high temperature anneal to improve the data retention of the memory cell. DRAM periphery development differs from high performance logic device scaling [23].

To support research, the Semiconductor Industry Association publishes the International Road map of Semiconductor (ITRS) [24]. The ITRS is a guide for the estimated necessities in a 15 years period in scaling. Guidelines of the ITRS 2009 for the DRAM regular peripheral devices are given in table 1.1. In the year 2013, the proposed gate stack is hafnium silicon oxynitride with a titanium nitride metal electrode for the PFET. To

Table 1.1: Requirements of the Peripheral DRAM Transistors after ITRS 2009

Device	NFET		PFET	
Year	2012	2013	2012	2013
Gate Dielectric	SiON	HfSiON	SiON	HfSiON
PFET Electrode	/	/	Polysilicon	TiN
Device Structure	planar	planar	planar	planar
V_{Supply} (V)	1V	1V	-1V	-1V
EOT* (nm)	2.5	2.3	/	/
L_{Gate} (nm)	65	55	/	/
I_{on} (μA/μm) V_{Gate}=V_{Drain}=V_{Supply}	400	400	-170	-175
V_{th} (V) V_{Drain}=±0.055V	0.4	0.4	-0.4	-0.4

*EOT- Equivalent electrical oxide thickness
All values given for regular peripheral devices at a temperature of 25°C.

reach the gate length target of 55nm, the diffusion behavior of the source/drain extension needs to be well controlled. Therefore, the PFET boron doping profile is regulated by a carbon implant, which reduces the transient enhanced diffusion.

A high-k gate stack, and additional carbon implantations are implemented in Qimonda technology to meet the ITRS targets of the year 2013. The main goal of this study is to understand how this process changes impact the leakage currents of the peripheral DRAM MOSFETs.

After a brief introduction, the most important leakage current mechanisms are presented in chapter 2. The applied electrical measurement methods, and simulation models are explained in chapter 3. The influence of the source/drain implants, such as carbon, on the leakage currents are studied in detail in chapter 4. Chapter 5 investigates the impact of high-k metal gate stack processing on transistor leakages. Especially the process of removing the high-k layer from the source/drain as well as the silicon nitride encapsulation of the gate stack are in the focus of chapter 5. A short summary concludes the thesis

2 Fundamentals of Leakage Currents in MOSFET Devices

Leakage currents significantly enhance power consumption of transistors in CMOS technology. Various off-currents can be differentiated depending on the bias conditions. Figure 2.1 gives an overview of the most important intrinsic leakages of a MOSFET. Extrinsic currents to the sidewall of the devices are not part of this investigation. The gate leakage (I_{Gate}) flows through the dielectric to the silicon. The subthreshold current (I_{SubVth}) arises between source and drain when the gate is switched off. Channel current ($I_{channel}$), source/drain current ($I_{S/D}$), source/drain extension current (I_{SDE}), and gate induced drain leakage (I_{GIDL}) flow from the corresponding space charge regions in silicon. All leakages will be explained in detail in this chapter.

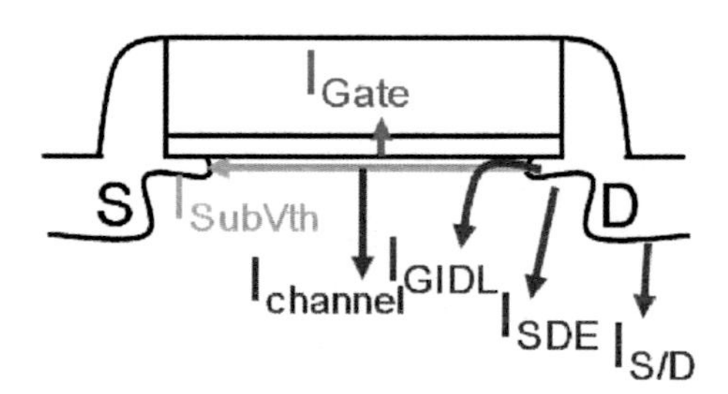

Figure 2.1: Overview of different leakage currents occurring in MOSFET transistors: Gate leakage (I_{Gate}), subthreshold current (I_{SubVth}), channel leakage ($I_{channel}$), source/drain current ($I_{S/D}$), source/drain extension current (I_{SDE}), and gate induced drain leakage (I_{GIDL}).

2.1 Leakage through Gate Oxide

The power dissipation due to leakage through a thin SiO_2 gate oxide increases energy consumption. High permittivity (high-k) materials are introduced in scaled transistors to further decrease the electrical thickness while maintaining a reliable physical thickness, and a low gate leakage [12].

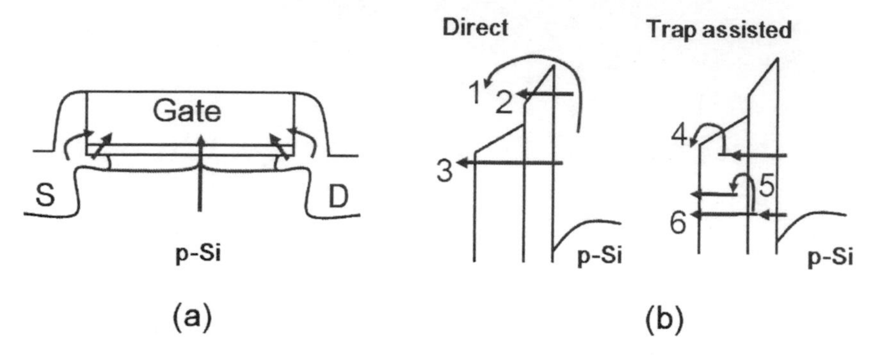

Figure 2.2: (a) Leakage current paths through the gate dielectric in a NFET are presented [25, 26].
(b) Different leakage current mechanisms through a high-k gate oxide are proposed: Schottky emission (1), Fowler Nordheim tunneling (2), direct tunneling (3), Frenkel Poole emission (4), hopping (5), and trap assisted tunneling (6) [27].

To compare transistors with different dielectrics the equivalent oxide thickness (EOT) is introduced. The EOT takes the ratio of the permittivity of the gate dielectric (ϵ_{ox}) to the permittivity of silicon oxide (ϵ_{SiO2}) into account (equation 2.1) [9]. If parasitics and quantum mechanical effects are neglected the equivalent oxide thickness can be extracted by capacitance voltage measurement, and is called capacitance equivalent oxide thickness (CET).

$$EOT = \frac{t_{ox}\epsilon_{SiO2}}{\epsilon_{ox}} \approx CET \tag{2.1}$$

Different gate leakage paths can be found for MOSFET devices [25]. One of them being an area dependent current through the gate oxide towards the channel or the bulk, the other one being a perimeter dependent current towards the source/drain (Fig. 2.2(a)) [25, 26].

Different leakage current mechanisms have been described in literature depending on the type of oxide, oxide thickness, and processing conditions (Fig. 2.2(b)). Each mechanism has a specific temperature and voltage dependences, and dominates at certain bias conditions [27]. The activation energy (E_a) depending on the electric field is calculated from temperature dependent measurements to determine the corresponding leakage mechanism (equation 2.2) [28].

$$E_a = \frac{d[ln(I_{Gate}/[A])]}{d[e/(k \cdot T)]} \tag{2.2}$$

2.1.1 Thermally Activated Gate Leakage

In the Schottky emission process, charge carriers are thermally enhanced over the barrier (mechanism (1) in Fig. 2.2(b)) [27]. An additional reduction of the barrier due to image force lowering takes place. Schottky emission current depends on the barrier height, and oxide thickness [27]. SiO_2 is a wide band gap material (E_g=9eV), so Schottky emission is suppressed [12].

For Frenkel Poole conduction, the charge carriers are thermally enhanced from a trap over the barrier (mechanism 4 in Fig. 2.2(b)) [27]. The barrier for the thermal enhancement is reduced by electrical interaction between the carrier and a charged defect [29]. The barrier height of the trap corresponds to the activation energy at zero field.

Thermally activated hopping from one defect to the next occurs in trap rich gate dielectrics [30] (mechanism (5) in Fig. 2.2(b)). A combined mechanism, of a thermal emission from a trap, and a tunneling over the remaining part of the barrier is also possible [31]. All of these mechanisms are preferentially found if the gate dielectric thickness is above 3nm.

2.1.2 Tunneling Leakage Through the Gate Dielectric

For the thin gate dielectrics approximately below 3nm in scaled MOSFETs, the tunneling current is the dominant [26]. The tunneling current is nearly independent of the temperature. Only small changes in the band gap, and the availability of carriers with the temperature lead to an activation energy below 0.15eV [32].

Trap assisted tunneling is an important leakage mechanism at low electric fields when the probability of tunneling through the complete barrier is low [33] (mechanism (6) in Fig. 2.2(b)). It can be observed for inelastic tunneling processes [33], and gets more important for high-k stacks with a SiO_2 interface layer [34]. At higher electric fields, the tunneling current is described by the Fowler Nordheim- or by the direct tunneling model [26].

The approximation of the Fowler Nordheim tunneling current is based on the Wenzel-Kramers Brillouin theory for the tunneling probability through a triangular barrier (mechanism (2) in Fig. 2.2(b)) [27]. The model assumes an elastic process of one charge carrier. A single effective mass, and the barrier height control the current [26]. No image force barrier lowering, or temperature dependent carrier density are taken into account [26]. The gate leakage (I_{Gate}) in this model depends on the oxide field (F_{ox}). The current is approximated by equation (2.3) [26]. A linear plot, called Fowler Nordheim graph, with a constant slope B_{FN} is obtained when the axes are scaled according to equation (2.3). The slope B_{FN} depends on the type of oxide with its specific barrier height (ϕ_b) and effective mass (m^*).

$$\frac{I_{Gate}}{A} = \frac{e^3 F_{ox}^2}{16\pi^2 \hbar \phi_b} exp\left(\frac{-B_{FN}}{F_{ox}}\right) = \frac{e^3 F_{ox}^2}{16\pi^2 \hbar \phi_b} exp\left(\frac{-4\sqrt{2m^*}\phi_b^{3/2}}{\hbar e F_{ox}}\right) \tag{2.3}$$

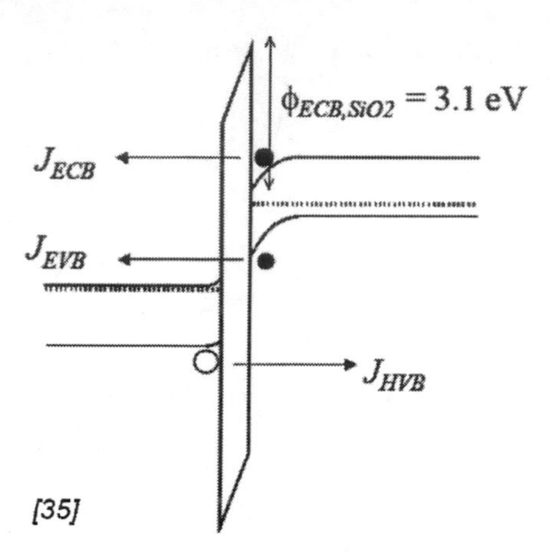

Figure 2.3: Different gate direct tunneling mechanisms are presented: Electron conduction band tunneling (ECB), electron valence band tunneling (EVB), and hole valence band tunneling (HVB) [35]. The barrier height ($\Phi_{ECB,SiO2}$) of ECB using silicon oxide as gate dielectric is given.

In transistors with thin gate dielectric below 3nm, tunneling through a trapezoidal barrier is no longer negligible (mechanism (3) in Fig. 2.2(b)) [26]. This process is described by the direct tunneling model. The slope B_{FN}, extracted by the Fowler Nordheim plot, is now dependent on the electric field. The Fowler Nordheim mechanism is a specific case of direct tunneling [26].

Depending on the gate voltage, different charge carriers dominate the transistor gate leakage (Fig. 2.3) [35]. Electron conduction band tunneling (ECB), and electron valence band tunneling (EVB) lead to the gate to bulk current during accumulation and inversion. The gate to channel current during inversion is due to hole valence band tunnel (HVB) for the PMOS, and ECB for the NMOS. The overlap leakage from the gate to the source or drain is governed by ECB [35].

2.2 Subthreshold Leakage from Source to Drain

The minority charge carrier density in the channel increases exponentially when the gate bias is increased from nearly 0V to the threshold voltage. These carriers lead to a significant diffusion current even at 0V. It is called subthreshold leakage (I_{SubVth}, Fig. 2.4) [3]. The current increases exponentially with the gate voltage. The slope is proportional to the body factor m. The body factor m is related to the ratio between the oxide capacitance, and the depletion layer capacitance. The body factor scaled with the temperature

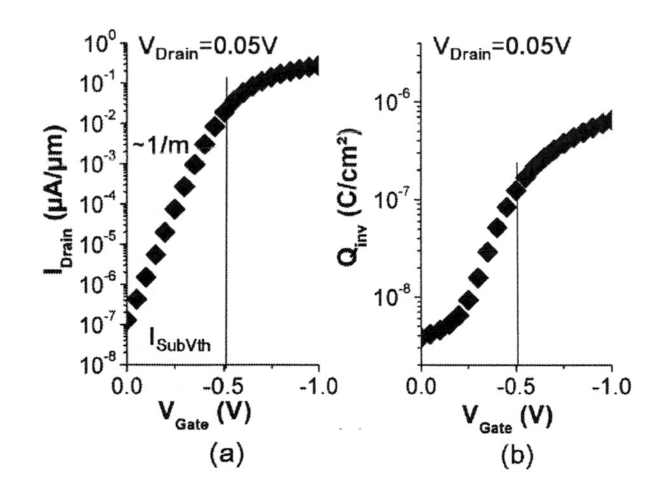

Figure 2.4: Subthreshold current, and inversion charge measured at PFETs with a gate length of 5μm, and a CET of 2.3nm at a temperature of 85°C.
(a) Subthreshold current depending on the gate voltage.
(b) Inversion charge versus gate voltage calculated from the transfer characteristic, and capacitance voltage measurements.

gives the subthreshold slope (S_{Vth}) [3].

The subthreshold current is approximated using equation (2.4) [3]. The leakage depends on the effective mobility (μ_{eff}), the oxide capacitance (C_{ox}), the width (W_{Gate}) and length (L_{Gate}) of the gate, the body factor (m), and the threshold voltage (V_{th}).

$$I_{Drain} = \mu_{eff} C_{ox} (m-1) \frac{W_{Gate}}{L_{Gate}} \left(\frac{kT}{e}\right)^2 \cdot exp\left[\frac{-e(V_{Gate}-V_{th})}{mkT}\right] \cdot \left(1 - exp\left[\frac{-eV_{Drain}}{kT}\right]\right) \tag{2.4}$$

In short channel devices, the subthreshold leakage is enhanced by the drain induced barrier lowering (DIBL) effect [36]. The potential barrier reduces when source- and drain depletion regions approach each other (Fig. 2.5). More charge carriers drift from source to drain [36]. The DIBL is described by a linear reduction of the threshold voltage with the drain bias (equation 2.5) [37]. The slope of this reduction is called DIBL parameter (λ). The DIBL parameter depends on the gate oxide (ϵ_{ox}=ϵ_{SiO2}), oxide thickness (CET), the gate length (L_{Gate}), the geometry correction factor (η), and the doping (m) [37].

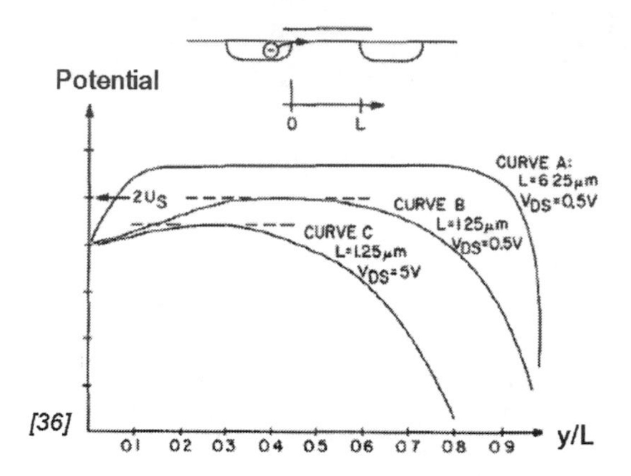

Figure 2.5: Surface potential distribution for constant gate voltages, and different channel length [36].

$$V_{th} = V_{th0} - \lambda \cdot V_{Drain} \qquad (2.5)$$
$$\lambda = \frac{\epsilon_{Si} \cdot CET \cdot m}{\epsilon_{SiO2} \cdot \eta \cdot L_{Gate}}$$

2.3 Leakage from Silicon Space Charge Regions to Transistor Bulk

Depending on the bias conditions, different types of leakage current flow to the transistor bulk (Fig. 2.1). The gate induced drain leakage at the transistor edge (I_{GIDL}), and leakage from the channel region ($I_{channel}$) occur due to the band bending at the dielectric to silicon interface. The source/drain current ($I_{S/D}$), and source/drain extension leakage (I_{SDE}) can be found at the pn-junctions.

2.3.1 Leakage Current Mechanisms

To explain the cause of the leakage currents, different mechanisms are suggested in literature (Fig. 2.6). They can be distinguished by their temperature, and voltage characteristics. The activation energy (E_a) is used to investigate the temperature dependence of the currents (equation 2.2). Depending on the leakage mechanism, the activation energy is related to the energy of the band gap or the trap energy.

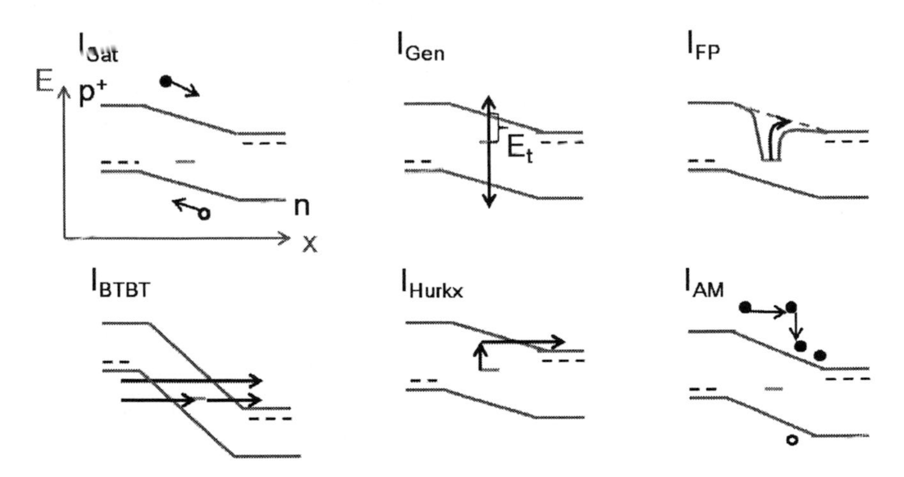

Figure 2.6: Leakage current mechanisms in the transistor depletion regions: Currents due to drift of minority carriers (I_{Sat}) [27], thermal generation (I_{Gen}) [27], Frenkel Poole transport (I_{FP}) [29], trap assisted tunneling (I_{Hurkx}) [38], band to band tunneling (I_{BTBT}) [38], and avalanche multiplication (I_{AM}) [39] are shown.

Saturation Current

If minority carriers diffuse into the depletion region, they drift across the junction (Fig. 2.6) [27]. The diffusion current of majority carriers is negligible under reverse bias [40]. This leads to a saturation current proportional to the doping concentration (N_{Dop}) at the low doped side of the junction [40]. The saturation current is independent of the electric field (equation 2.6) [27]. The temperature (T) dependence of the current is given by the diffusion coefficient (D_{Diff}), diffusion lifetime (τ_{Diff}), and the intrinsic carrier concentration (n_i). The activation energy calculated from the temperature dependent curves is approximately equal to the band gap energy (E_g).

$$\begin{aligned} I_{Sat} &= Ae\frac{n_i^2}{N_{Dop}}\sqrt{\frac{D_{Diff}}{\tau_{Diff}}} \\ &\propto T^{3+\gamma/2}\cdot exp\left(\frac{-E_g}{kT}\right) \\ E_a &= E_g \end{aligned} \tag{2.6}$$

Generation Current

Charge carriers emitted from traps in the band gap lead to an additional leakage current (Fig. 2.6) [27]. Most models available in literature assume one dominant trap. The Shockley Read Hall (SRH) theory describes a thermally activated carrier generation [41, 42]. The generation current increases with depletion width (W) which leads to an increase of leakage with the reverse bias (equation 2.7) [27, 43]. The temperature dependence of the leakage is caused by the intrinsic charge carrier density, and the generation lifetime (τ_{Gen}). The generation lifetime depends on the defect density (N), the capture cross section (σ) for the different charge carriers, and the trap energy (E_t) (equation 2.7).

$$\begin{aligned} I_{Gen} &= Ae\frac{n_i(W - W(0V))}{\tau_{Gen}} \\ \tau_{Gen} &= \frac{\sigma_h \sigma_e v_{th} N}{\sigma_n exp((E_t - E_i)/kT) + \sigma_p exp((E_i - E_t)/kT)} \end{aligned} \tag{2.7}$$

For mid gap traps, the carrier generation probability is high, and the generation lifetime is low. The generation lifetime is a slowly varying function of the temperature for traps where the defect energy equals half of the silicon band gap energy (E_i=$E_g/2$) [27]. The activation energy calculated from the current voltage curves equals then half of the energy of the silicon band gap energy due to the intrinsic charge carrier density. If the trap is positioned below or above mid gap, the generation lifetime causes the temperature dependence (equation 2.8) [27, 43]. The difference between the trap energy, and mid gap ($|E_t$-$E_i|$) is determined from the activation energy (equation 2.8).

$$\begin{aligned} I_{Gen} &\propto \frac{T^{3+\gamma/2} exp(-E_g/2kT)}{\sigma_n cosh((E_t - E_i)/kT)} \\ E_a &= E_t \ (E_t \geq E_i) \\ E_a &= E_g - E_t \ (E_t < E_i) \end{aligned} \tag{2.8}$$

Frenkel Poole Mechanism

Electric interaction between a charge carrier and a charged trap effects the band bending, and leads to a reduction of the barrier height for thermal generation (Fig. 2.6) [29]. The reduction in barrier height (ΔE_a) is dependent on the electric field (F_{eff}). The effect is modeled by equation (2.9) [29].

$$\Delta E_a = \sqrt{\frac{e^3 F_{eff}}{\pi \epsilon_{Si}}} \tag{2.9}$$
$$I_{FP} = I_{Gen} B_{FP} F_{eff} \cdot exp\left(\sqrt{\frac{e^3 F_{eff}}{\pi \epsilon_{Si}}} \frac{1}{kT}\right)$$

The Frenkel Poole effect leads to an increased generation current, and an electric field dependent activation energy. A decrease of activation energy with the square root of the electric field of -22μeVm$^{1/2}$V$^{-1/2}$ for a single charged defect is found (equation 2.10) [29].

$$I_{FP} \propto T^{3+\gamma/2} \cdot exp\left(\frac{E_t}{kT}\right) \cdot exp\left(\sqrt{\frac{-e^3 F_{eff}}{\pi \epsilon_{Si}}} \frac{1}{kT}\right) \tag{2.10}$$
$$E_a = E_t - \sqrt{\frac{e F_{eff}}{\pi \epsilon_{Si}}}$$

Tunneling Current

Band to band tunneling, or trap assisted tunneling takes place if there are occupied energy levels at one side of the depletion zone, and unoccupied levels at the other side (Fig. 2.6) [44]. The tunneling probability through a triangular barrier can be calculated by the Wenzel Kramers Brillouin approximation [27]. It depends on the barrier height ($E_g/2$), the electric field (F_{eff}), and the effective mass (m^*) [44]. The tunneling current is almost temperature independent. The activation energy of the tunneling process is below 0.15eV [32]. The small temperature dependence is due to the change of the energy gap, and phonon assistance in the indirect tunneling process [44]. Trap assisted tunneling, and direct tunneling have the same emission probability [45].

Groups around Hurkx [38, 43], and Schenk [46] developed a model for the tunneling current. In the Hurkx model for band to band tunneling in silicon diodes, the leakage current is approximated using equation (2.11) [43].

$$I_{BTBT} = B_{BTBT} V \left(\frac{F_{eff}}{F_0}\right)^{3/2} A \; exp\left(\frac{-F_0}{F_{eff}}\right) \tag{2.11}$$
$$F_0 = \frac{\pi \sqrt{m^*} E_g^{3/2}}{2\sqrt{2}\hbar} \; (direct\ tunneling)$$
$$F_0 = 1.9 \cdot 10^7 V/cm \; (Si\ 25°C)$$

Trap Assisted Tunneling with Phonon Interaction (Hurkx)

For the trap assisted tunneling, a two step mechanism is proposed by Hurkx [38, 43], and Schenk [46] (Fig. 2.6). First, the charge carrier is thermally enhanced, and then it tunnels through the remaining part of the barrier. The thermal generation current is enhanced by a tunneling factor (B_{Hurkx}).

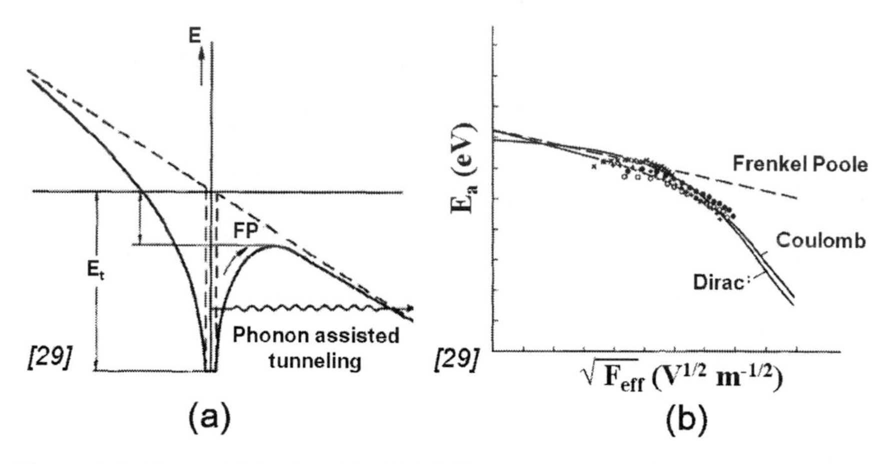

Figure 2.7: The model developed by M.J.J. Theunissen, and F.J. List is presented [29]. (a) Schematic of the band diagram including thermal carrier generation enhanced by the Frenkel Poole effect (FP), and trap assisted tunneling with phonon interaction from a Dirac- or a Coulomb well (Hurkx).
(b) The change of activation energy with the electric field for the models from publication [29] is shown.

The tunneling factor depends on the effective mass (m^*), the depletion width (W), the effective electrical field (F_{eff}), and the temperature (T). The current (I_{Hurkx}) of the trap assisted Hurkx model is analytically approximated by equation (2.12) [43].

$$\begin{aligned} I_{Hurkx} &= I_{Gen} \cdot B_{Hurkx} \\ &= I_{Gen} \cdot \sqrt{3\pi}\frac{F_\Gamma}{F_{eff}}\left(exp\left[\frac{F_{eff}}{F_\Gamma}\right]^2 - exp\left[\frac{F_{eff} \cdot W(0V)}{F_\Gamma \cdot W}\right]^2\right) \\ F_\Gamma &= \frac{\sqrt{24m^* \cdot (kT)^3}}{e\hbar} \end{aligned} \tag{2.12}$$

The activation energy (E_a) is dependent on the bias. E_a decreases more rapidly with the voltage than in the Frenkel Poole theory (Fig. 2.7) [29]. If the slope of the activation energy with the square root of the electric field is about an order of magnitude lower than -0.22μeVm$^{1/2}$V$^{-1/2}$ [47], this can be considered as an indication for the trap assisted tunneling mechanism with phonon interaction. The Frenkel Poole effect can be neglected in this case (Fig. 2.7) [38]. The trap energy (E_t) of the generation current is extracted from this leakage current by taking the tunneling factor (B_{Hurkx}) into account [48].

Avalanche Multiplication

The charge carriers in the depletion region of the junction are accelerated in the electric field. If the kinetic energy of the charge carriers is high enough to generate additional electron hole pairs an increase in leakage current is measured (Fig. 2.6) [39]. This process is called avalanche multiplication.

2.3.2 Perimeter Leakage from Source/Drain to Bulk

This part of the section focuses on the location of the different depletion regions where the perimeter currents: Source/drain leakage, source/drain extension leakage, and gate induced drain leakage, are generated (Fig. 2.8).

Source/Drain and Source Drain Extension Leakage

When gate, source, and drain are shorted, only source/drain extension (I_{SDE}) leakage, and source/drain leakage ($I_{S/D}$) are induced at the MOSFET perimeter (Fig. 2.8(a)). These two leakage currents are distinguished because they occur in regions with different doping gradients, and defect concentrations.

The source/drain to bulk current is a diode leakage (Fig. 2.8(a)). It is generated deep in the silicon bulk. Therefore, the effect of the implantation damage caused during the source/drain formation on this leakage current is reduced [48]. The vertical doping profile determines the electric field in the source/drain junction (Fig. 2.9(b)).

The source/drain extension leakage is generated in the depletion region with the peak implantation damage from the source/drain formation (Fig. 2.8(a)). Also, the doping concentration, and therefore the electric field in the junction is changing from the silicon surface downwards (Fig. 2.9(a)).

All current mechanisms mentioned in section 2.3.1 can theoretically dominate the perimeter leakage. For the devices used in this study, experimental results show that the S/D-, and the SDE currents are dominated by the same mechanism.

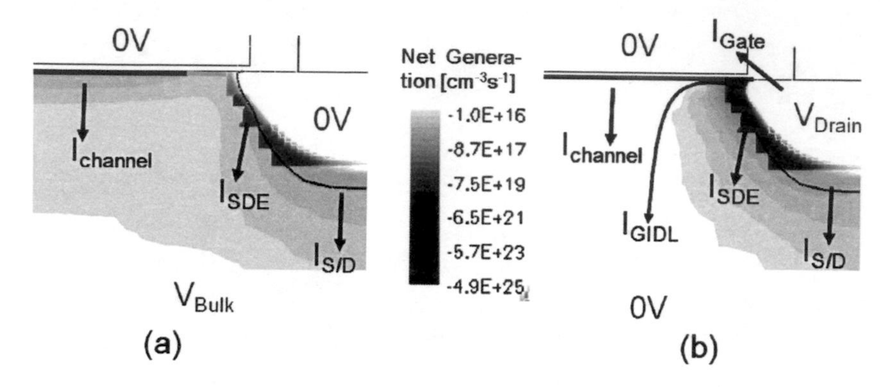

Figure 2.8: TCAD simulations of the leakage paths in a PFET under off conditions when:
(a) The gate, source, and drain are shorted.
(b) The source/drain bias is ramped to form a deep depletion.

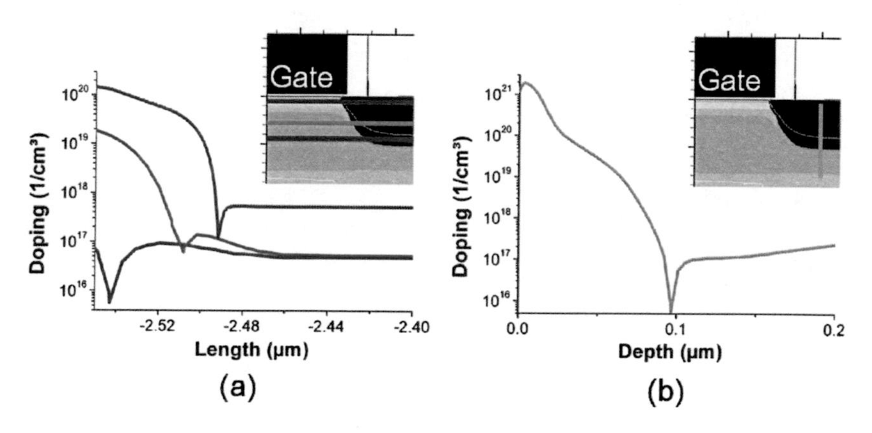

Figure 2.9: Simulated PFET doping profiles at:
(a) Different horizontal cut lines.
(b) A vertical cut line through the drain.

Gate Induced Drain Leakage

The band bending in source/drain overlap region leads to a deep depletion when the gate is biased to a depletion- or accumulation voltage (Fig. 2.10) [49]. An additional current occurs in this region, called gate induced drain leakage (I_{GIDL}) (Fig. 2.11, Fig. 2.8(b)).

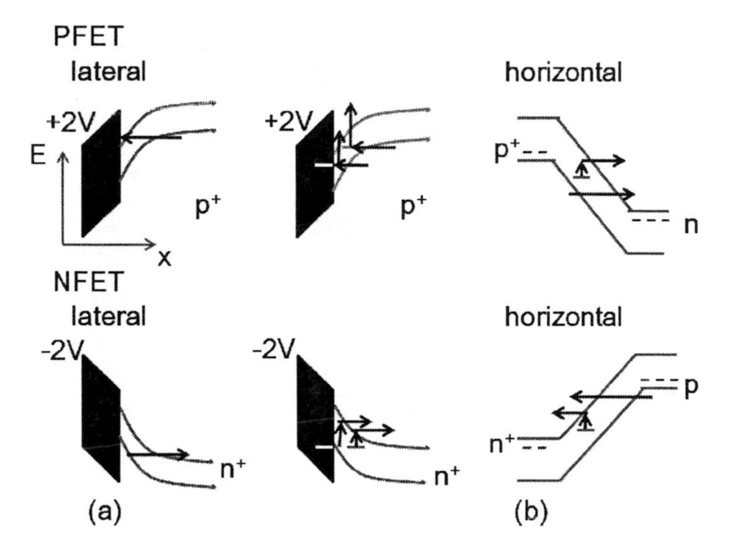

Figure 2.10: Schematic band diagram of PFET and NFET GIDL due to the:
(a) Band bending at the gate oxide to silicon interface [50, 51].
(b) Depletion of the source/drain extension junction [50, 51].

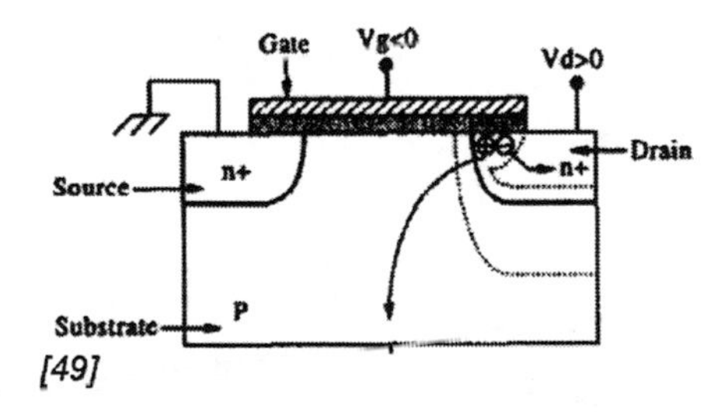

Figure 2.11: Deep depletion forming in an NFET overlap region leading to GIDL after publication [49].

Figure 2.10 shows the band bending, and the possible leakage current mechanisms of the GIDL. Direct tunneling, and trap assisted tunneling via interface defects or bulk defects close to the silicon interface are often proposed leakage mechanisms [51, 52]. The GIDL current of the NFET devices is relatively insensitive to the interface traps at the gate edge. Only interface traps close to the silicon valence band are occupied, and contribute to the generation of the NFET GIDL current (Fig. 2.10) [50]. In case of PFETs, electrons tunnel into the interface traps, and are thermally enhanced [50]. The GIDL current is modeled in literature with the numerical approximation made by Hurkx et al. for band to band tunneling, and trap assisted tunneling [38, 51].

2.3.3 Generation Leakage from Channel Region to Bulk

This section focuses on the area dependent leakage that is generated in the depletion region beneath the gate (Fig. 2.8). It will be called channel leakage in this study. The conditions are similar to the deep depletion leading to the GIDL current. But, the band bending in the channel area is lower compared to the highly doped overlap region.

The channel leakage can be also approximated with model proposed by Hurkx et al. [38]. Due to the smaller band bending, the influence of the tunneling enhancement factor is reduced. When the gate is biased in depletion, the channel leakage ($I_{channel}$) is generated at interface defects, or bulk defects close to the interface (Fig. 2.12). The NFET channel current shows a stronger dependence on the voltage than the PFET channel current because electrons tunnel into the interface traps. For the PFET channel current, the thermal generation mechanism is dominant. Only interface traps close to the valance band edge contribute to the trap assisted tunneling channel current of the PFETs. Under inversion conditions, the channel current is generated at silicon bulk traps beneath the inversion charge carrier maximum.

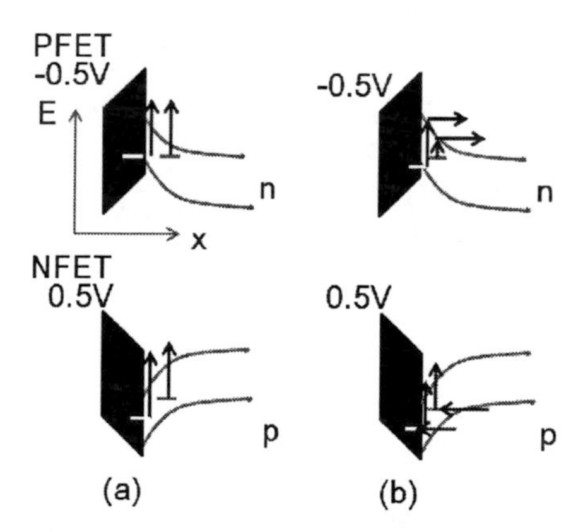

Figure 2.12: Schematic band diagram of the PFET and NFET channel leakage due to:
(a) Thermal generation.
(b) Trap assisted tunneling with phonon interaction.

3 Methodology

The electrical measurements are performed on fully processed 300mm wafers using the Qimonda 46nm and 65nm technology. To give a good signal to noise ratio for the leakage current-, and capacitance voltage measurements an array of MOSFET's with an overall area of at least $5000\mu m^2$ is used. The two test structures, called MultiFET and Overlap, are presented in Figure 3.1. The MultiFET is a four terminal structure with source, drain, gate, and bulk. In addition to the transistor array, the structure contains a single MOSFET with the same length. Due to its proximity, and similar layout this single transistor in the MultiFET structure is assumed to have the same offsets. The single device is used to measure the MOSFET transfer characteristics. In the overlap test structure, source and drain are shorted. The test structures differ in area to perimeter ratios (table A.1, A.2).

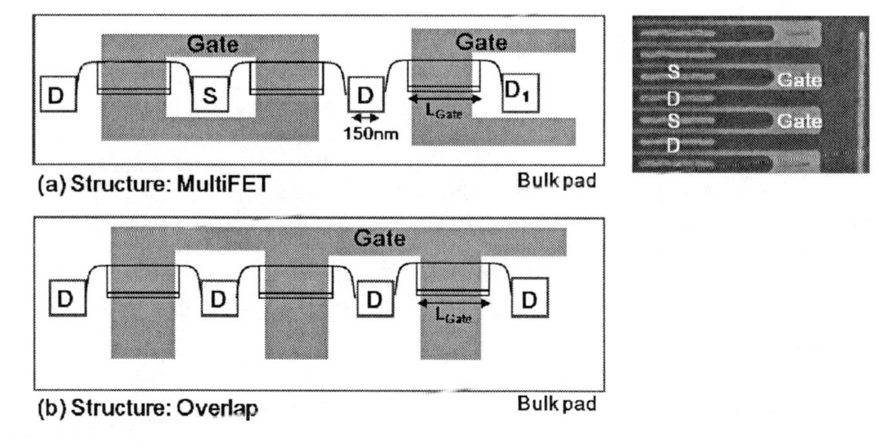

Figure 3.1: Schematics and overview of the:
(a) MultiFET test structure (scanning electron microscopy [53]).
(b) Overlap test structure.

A Cascade Summit 12000 probe system is used to handle the wafer. The test structures are contacted with single probes which are shielded towards the tip for low current measurements. Measurements are done in a temperature range of 0°C to 200°C. The temperature calibration is done with a thermal couple (table A.3). The temperature variability across the chuck is $\pm 1°C$. To avoid the condensation of water on the wafers, the samples are heated up before measurement. This procedure minimizes the moisture concentration in the probe system. In addition, a flux of dry air runs across the wafer.

A Keithley 4200 parameter analyzer with sense measurement unit and preamplifier, capacitance voltage unit, oscilloscope, and pulse generator is used to do the electrical analysis. With the equipment an accuracy of 100fA for current voltage (IV) at 25°C and a 20pF resolution for capacitance voltage measurements (CV) at 500kHz is reached (see also section A.3 of the appendix). The test setup is accurate for frequencies from 40kHz to 2MHz (Fig. A.3). The dissipation factor of the the CV measurements is below one [28]. Pulse generator, and oscilloscope are shorted across the drain metalization of the MOSFETs to test the accuracy of the pulse generator. Pulses with rise- and fall times down to 100ns can reliably be used for charge pumping measurements (Fig. A.4).

At least five dies are tested across each wafer for the manual measurements. The basic electrical characteristics of these dies are within the 0.25 to 0.75 quantile of the full wafer investigation by Qimonda. The failure bar, given for the measurements, is three times the standard deviation. To verify the interpretation of the electrical measurement, the TCAD Sentaurus device simulator [54], and the ngspice simulator [55] are used.

3.1 Current Voltage Measurement

Electrical measurements are used in industry to statistically evaluate the devices after processing. These tests are relatively fast, and less expensive than physical characterization methods. Extensive current voltage measurements can give indirect information on structural changes in the devices for different process variations. The first part of the section explains how the most important working parameters determining the transistor performance are extracted. The second part shows how the different off-currents which enhance the power consumption are separated. At the end the gated diode measurement, an efficient way to study the generation of charge carriers in different regions of a MOSFET [28], is introduced.

3.1.1 Basic Current Voltage Characteristics

Transfer characteristics, drain current (I_{Drain}) versus gate voltage (V_{Gate}), are measured to analyze the main electrical MOSFET parameters. Figure 3.2 shows the parameters that are extracted from the measurement. All curves are measured at a temperature (T) of 85°C.

In this study, the threshold voltage (V_{th}) is defined using the transfer curve at the transition point from weak inversion (subthreshold current) to strong inversion (linear MOSFET current). The threshold voltage is determined using the Y-Function method (equation

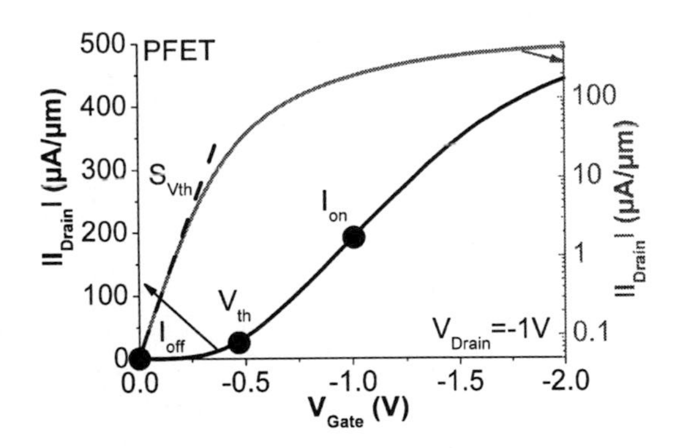

Figure 3.2: Transfer characteristic of a short channel PFET (L_{Gate}=65nm) with an CET of 2.3nm at 85°C. The on-current (I_{on}), off-current (I_{off}), threshold voltage (V_{th}), and subthreshold slope (S_{Vth}) are extracted from the measured curve.

3.1) [56]. This method allows to extract the V_{th} independent of the series resistance [56]. The correlation coefficient of the linear regression is above 0.95 in the linear voltage regime of the transistors. The on-current (I_{on}) is measured biasing gate, and drain to ±1.0V.

$$\begin{aligned} Y &= \frac{I_{Drain}}{\sqrt{dI_{Drain}/dV_{Gate}}} = \sqrt{\frac{W_{Gate}\mu_{eff}C_{ox}V_{Drain}}{L_{Gate}}}(V_{Gate} - V_{th}) \\ V_{th} &= V_{Gate}\,(Y = 0) \end{aligned} \tag{3.1}$$

The subthreshold slope (S_{Vth}), and body factor (m) are determined from the exponential increase of the drain current with the gate bias below threshold voltage (equation 3.2) [3]. The ideal S_{Vth} is 60mV/dec at 25°C. In this case the body factor is directly related to the ratio of depletion layer, and oxide capacitance [3]. The drain induced barrier lowering parameter (λ) is calculated from the threshold voltage shift of the transfer curves with V_{Drain} (equation 3.3) [57].

$$S_{Vth} = log\left(\frac{dV_{Gate}}{dI_{Drain}}\right) = ln(10)\cdot\frac{kTm}{e} \tag{3.2}$$

$$\lambda = \frac{V_{th}(V_{Drain} = \pm 0.05V) - V_{th}(V_{Drain} = \pm 1V)}{1V - 0.05V} \tag{3.3}$$

3.1.2 Separation of Leakage Path's

In order to analyze the different leakage paths, the off-currents are separated (Fig. 3.3). In the following, the method of separation is explained.

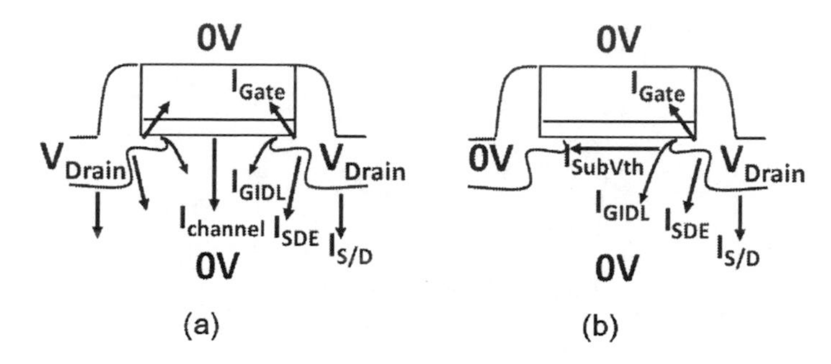

Figure 3.3: Schematic of the leakage current paths occurring in off-state (V_{Gate}=0V) when:
(a) V_{Source}=V_{Drain}.
(b) V_{Source}=0V.

Gate Leakage

Different leakage currents through the gate oxide (I_{Gate}) occur when the gate is biased (Fig. 2.2(a)). The area dependent gate to channel current is drawn towards source, and drain. It is dominant when the transistor is biased in the subthreshold or inversion regime. Additionally, an area dependent current between gate and bulk occurs. It can be measured at the bulk contact when the gate is biased. The perimeter dependent gate overlap leakage flows directly towards source, and drain. The gate overlap leakage is measured when the gate is biased to 0V, and the source/drain voltage is ramped (Fig. 3.3(a)).

Subthreshold Leakage

The subthreshold current (I_{SubVth}) depends on the potential difference between source and drain. The subthreshold current is measured directly at the source contact (Fig. 3.3(b)).

Source/Drain- , Source/Drain Extension- and Channel Leakage

The area dependent leakage from the channel region (channel leakage), and the perimeter dependent S/D- and SDE current occur when the gate is shorted with source, and drain (Fig. 3.4(a)). The area component (J_{Area}), and perimeter component ($J_{Perimeter}$) of the leakage are separated measuring two structures with different area to perimeter ratio (equation 3.4) [58].

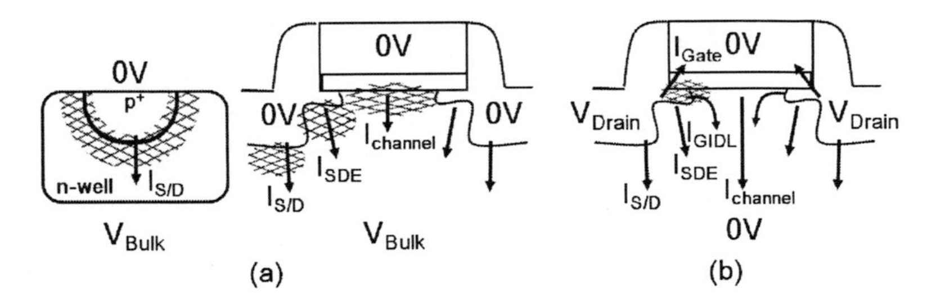

Figure 3.4: Leakage current flow under off conditions when the:
(a) Gate, source, and drain are shorted.
(b) Source/drain bias is ramped to form a deep depletion.

$$
\begin{aligned}
I_{Total1} &= J_{Area} \cdot A_1 + J_{Perimeter} \cdot P_1 \\
I_{Total2} &= J_{Area} \cdot A_2 + J_{Perimeter} \cdot P_2
\end{aligned}
\tag{3.4}
$$

Figure 3.5 shows an example calculation for a long channel PFET with an CET of 5.2nm. The S/D to bulk leakage current is measured on a separate diode structure (Fig. 3.4(a)). The current from the silicon to the shallow trench isolation can not be neglected when a diode structure is used for S/D current investigation. The area dependent diode leakage component has to be determined by equation (3.4). The SDE to bulk leakage current is determined subtracting the measured diode leakage from the silicon perimeter current.

The S/D-, SDE- and channel geometries of the MOSFETs are approximated from the test structure layout (Fig. 3.6). An example calculation for a MultiFET device with a gate length of 5μm is shown in equation (3.5).

$$
\begin{aligned}
A_{S/D} &\approx 200 \cdot 2 \cdot 5\mu m \cdot 0.2\mu m = 400\mu m^2 \\
A_{SDE} &\approx 200 \cdot 2 \cdot 5\mu m \cdot 0.1\mu m = 200\mu m^2 \\
A_{channel} &\approx 200 \cdot 5\mu m \cdot 5\mu m = 5000\mu m^2
\end{aligned}
\tag{3.5}
$$

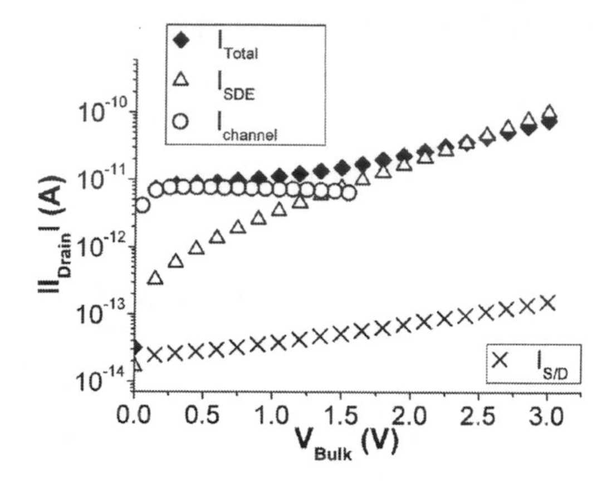

Figure 3.5: S/D-, SDE-, and channel leakage determined at a PFET for the sample set with different junction implants at 85°C. The measured PFET has a gate length and width of 5μm and a CET of 5.2nm.

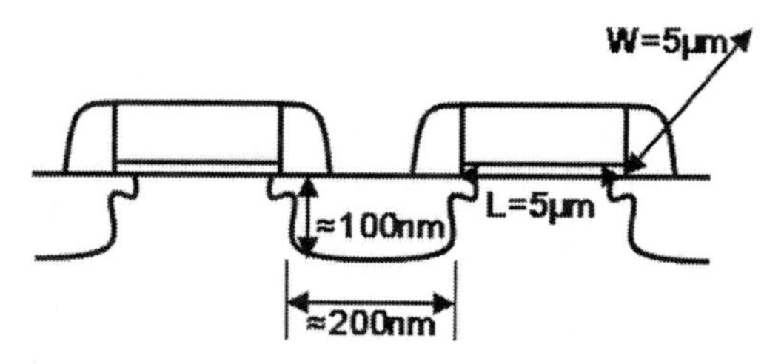

Figure 3.6: Test structure design of the MultiFET with L_{Gate}=5μm transistors for the current voltage measurements.

Gate Induced Drain Leakage

The gate induced drain leakage (GIDL) occurs when the gate is biased to depletion or accumulation, and the source/drain are ramped to reverse voltages (Fig. 3.4(b)). In addition, gate overlap leakage, channel current, S/D current, and SDE current occur. The gate overlap leakage is measured on the gate terminal. The channel-, S/D-, and SDE leakage are measured when the bulk is biased (Fig. 3.4(a)). The GIDL component is determined by equation (3.6) [51]. There is a small error in this procedure due to the fact that the generation in the channel is reduced when source and drain are ramped compared to the generation when the bulk is biased. Due to this reduction, there is a variation in the GIDL calculation at a drain bias below ±0.5V (Fig. 3.7). To avoid this uncertainty, GIDL IV characteristics will be analyzed at medium and high voltages which corresponds to a bias range from ±0.5V to ±3V for the MOSFETs in this study.

$$I_{GIDL} = I_{Drain} - I_{Gate} - I_{S/D} - I_{SDE} - I_{channel} \tag{3.6}$$

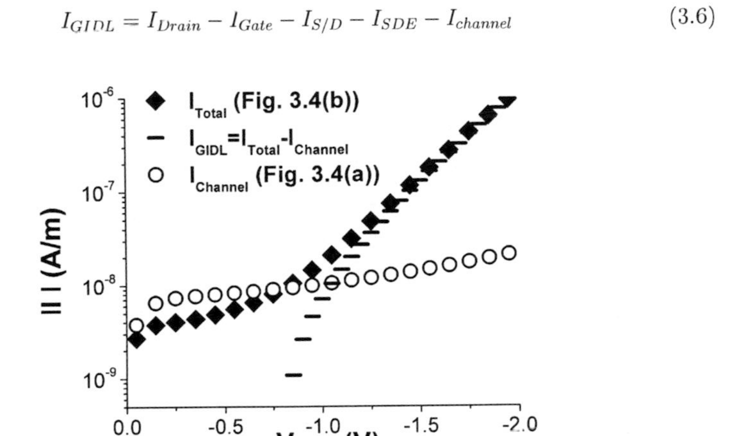

Figure 3.7: Channel current when the bulk is biased (Fig. 3.4(a)) compared to the total current, and the calculated GIDL current when a deep depletion is formed (Fig. 3.4(b)). Measurements are done on PFETs with L_{Gate}=5μm and a CET of 5.2nm at 85°C.

3.1.3 Gated Diode Measurement

The gated diode measurement is used to study the generation of charge carriers in different regions of a MOSFET [28]. The method was originally developed for gated diode structures [59] and was then adapted for transistor structures.

For the measurement the gate voltage is swept from inversion to accumulation, and a common source/drain reverse voltage is applied. Under inversion, the current is generated at the SDE depletion region, the S/D depletion region, and field induced depletion region beneath the inversion layer (Fig. 3.8(a)) [60]. During depletion, the current is mainly generated at the interface traps, and defects close to the silicon to gate oxide interface (Fig. 3.8(b)). The current measured in depletion reaches a maximum [60]. When the gate is biased to accumulation, a deep depletion forms at the source/drain extension to gate overlap [60]. The GIDL is the main leakage in accumulation. It shows a strong increase with the gate voltage (Fig. 3.8(c)).

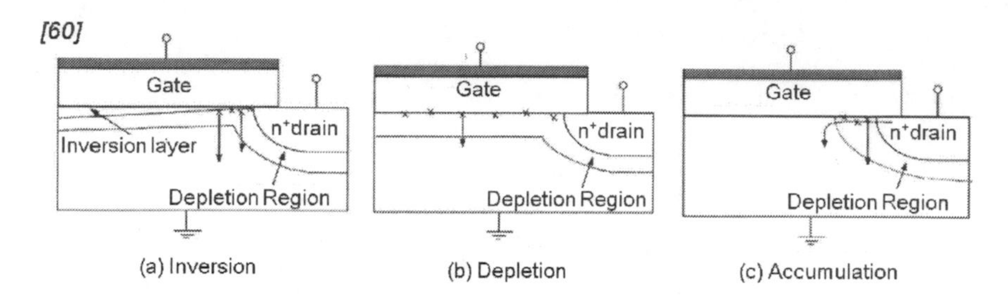

Figure 3.8: Schematics of the currents flowing to the bulk during a gated diode measurements under different gate voltages [60]. Interface defects are marked with crosses in this figure.

The measured leakage currents are simulated with TCAD device software. To validate the assumption of the TCAD model, measured, and simulated gated diode characteristics are compared. Figure 3.9 presents a simulated gate diode current for one of the devices of this study.

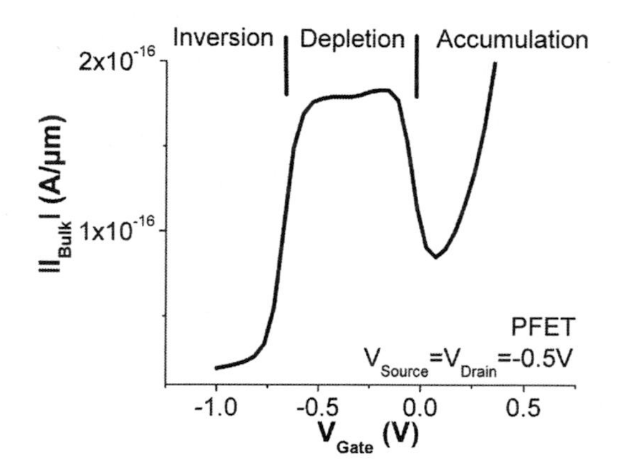

Figure 3.9: Current voltage characteristic of a gated diode measurement simulated for a long channel PFET device with a CET of 5.2nm and a gate length of 5μm. The measurement temperature is 25°C.

3.2 Capacitance Voltage Measurement

From capacitance voltage (CV) measurements, important diode- and MOSFET parameters are extracted. All CV measurements are done at 500kHz, and a signal amplitude

50mV. The measurement temperature is 85°C which is also used to analyze the transfer curves.

From the junction capacitance (C) voltage (V_{Diode}) characteristics, the effective electric field (F_{eff}), and depletion width (W) are determined by fitting the measurements to the theoretical curve (equation 3.7) [43]. The free parameter r, and the parameter V_{int} which is related to the build in potential of the junction, are used in the procedure. Figure 3.10 shows the measurement, and the corresponding fit for one diode of this study.

$$\begin{aligned} C &= \frac{C(0V)}{(1 - V_{Diode}/V_{int})^r} \\ W &= \frac{\epsilon_{Si}}{C} \\ F_{eff} &= \frac{V_{int}}{(1-r)W(0V)} \cdot (1 - V_{Diode}/V_{int})^{1-r} \end{aligned} \tag{3.7}$$

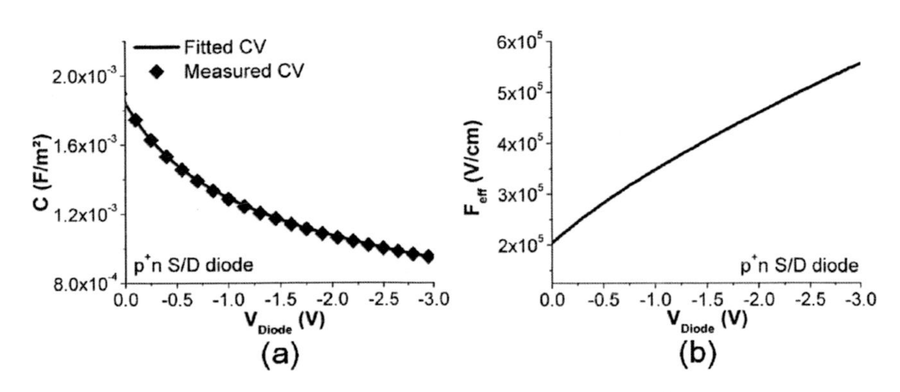

Figure 3.10: Measured and fitted CV characteristics for a p^+n diode structure. The parameters used for fitting are r=0.4, and V_{int}=0.7.
(a) Comparison of measured, and fitted CV curve using equation (3.7). The measurement is done at 85°C and 500kHz.
(b) Effective electric field calculated from the measured CV curve using equation (3.7).

Figure 3.11(a) exhibits the change of MOSFET capacitance with gate voltage. When the gate is biased to accumulation, the majority carriers gather at the gate dielectric to silicon interface, and only the oxide capacitance (C_{ox}) can be measured [3]. At the flat band voltage no band bending occurs. When the gate is biased to depletion, a space charge region in the silicon forms. The depletion width changes with the gate voltage. The total capacitance can be modeled as gate oxide- and silicon depletion capacitance in series. The total capacitance reduces with voltage until the maximum depletion width depending on the doping is reached [3]. At threshold voltage, the capacitance measured between gate and bulk drops to zero in the ideal case. The capacitance value between gate and source/drain increases rapidly up to the oxide capacitance when the gate is biased towards inversion. If a polysilicon gate is used, the capacitance drops slightly at high inversion voltages due to the polysilicon depletion capacitance in series [3].

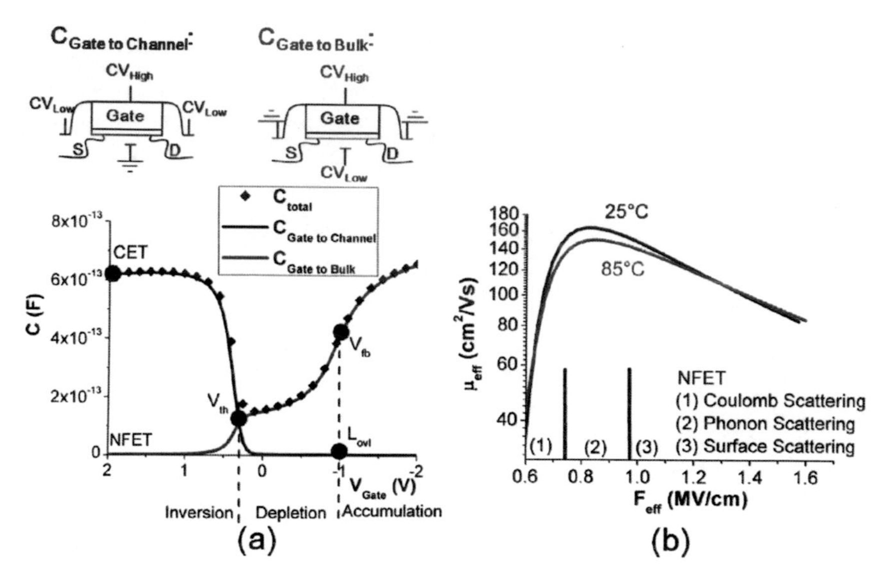

Figure 3.11: Mobility determination by split CV measurements at 5μm long NFETs with a CET of 2.5nm.
(a) Measurement setup, and split CV results at 85°C are presented. The equivalent oxide thickness (CET), flatband- and threshold voltage (V_{fb}, V_{th}), and the overlap length (L_{ov}) are determined from the CV curves.
(b) Mobility calculated from split CV measurements, and NFET transfer curves at different temperatures.

The capacitance equivalent oxide thickness (CET), flatband voltage (V_{fb}), threshold voltage (V_{th}), and overlap length (L_{ov}) are extracted from the CV curves (Fig. 3.11 (a)). Additionally the inversion- (Q_i) and accumulation charge (Q_{acc}), the gate oxide field (F_{ox}), and the mobility (μ_{eff}) are determined.

The flatband voltage is calculated from the inflection point of the gate to bulk CV curve [3]. Similarly the threshold voltage is extracted at the inflection point of the gate to source/drain measurement. The CET is calculated from the maximum accumulation capacitance (C_{ox}, equation 3.8) [12]. The CET is directly related to the physical oxide thickness (t_{ox}) according to equation (2.1).

$$CET = \frac{\epsilon_0 \epsilon_{SiO2} A}{C_{ox}} \tag{3.8}$$

Under non ideal conditions, especially for short channel devices, the capacitance does not drop to zero. Parasitic capacitances occur at the overlap region of the transistors (Fig 3.12) [3], and at the supply lines. In this study, the parasitics are nearly constant at ±2V. The CV curve is corrected by subtracting the offset values. The resulting gate to source/drain capacitance at flatband voltage is the inner overlap capacitance (C_{ov1}, Fig 3.12) [3]. The overlap length is then calculated by equation (3.9) [3]. This method enables to directly measure differences in overlap length for the various processing conditions at the same test structure.

$$L_{ov} = \frac{C_{Gate\ to\ Source/Drain}(V_{fb})}{2 C_{ox} W_{Gate}} \tag{3.9}$$

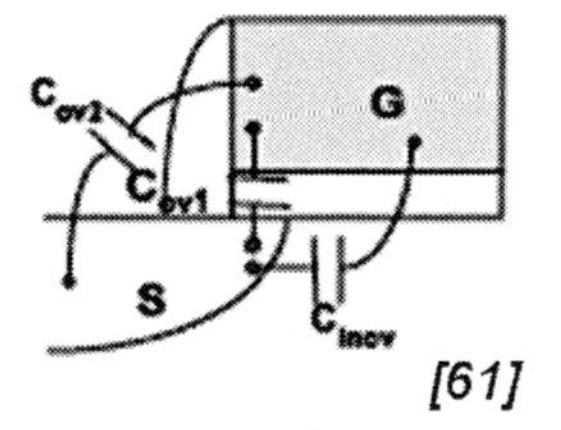

Figure 3.12: Parasitic overlap capacitances (C_{ov}) occurring in a MOSFET after publication [61].

The inversion charge and accumulation charge is obtained integrating the CV curves (equation 3.10) [28]. The effective field in silicon is determined from the charges. The effective oxide field can be calculated if the relative permittivity of the oxide is known [3, 27].

The mobility is directly related to the on-current. The mobility is calculated from the split CV - and the transfer curves for long channel devices (equation 3.11) [28]. The mobility is limited by coulomb scattering at charged centers at the interface in the low field regime, by phonon scattering at medium fields, and by scattering at surface roughness such as interface defects at high fields (Fig 3.11(b)) [62, 63].

$$
\begin{aligned}
Q_i &= \int_{\pm\infty}^{\mp V_{Gate}} C_{Gate\ to\ Source/Drain} dV_{Gate} \\
Q_{acc} &= \int_{\pm V_{Gate}}^{\mp\infty} C_{Gate\ to\ Bulk} dV_{Gate} \\
F_{Si} &= \frac{bQ_i + Q_{acc}}{\epsilon_{Si}} = \frac{F_{ox}\epsilon_{ox}}{\epsilon_{Si}} \\
b &= 1/2\ (electrons);\ b = 1/3\ (holes)
\end{aligned}
\tag{3.10}
$$

$$
\mu_{eff} = \left[\frac{dI_{Drain}}{dV_{Drain}}\right] \cdot \frac{L_{Gate}}{W_{Gate}Q_i}
\tag{3.11}
$$

3.3 Charge Pumping Measurement

Charge pumping (CP) measurements are used to determine the trapping characteristics at the silicon to gate oxide interface. A pulse between accumulation and inversion is applied at the gate, and the current is measured at the bulk (Fig. 3.13(a)) [64]. Source and drain are shorted, and biased towards small forward or reverse current. The CP characteristics of a PFET will be described in the following text, but all equations hold true for the NFET, respectively.

Figure 3.13(b) shows the change of voltage at the source/drain, and the bulk during the gate pulse. The voltage change is directly related to the injected charge leading to a current flow at the terminals. When the gate is biased to accumulation, electrons are gathered at the channel. Some of the electrons are trapped. During the transition to inversion the electrons flow towards the bulk, leading to the bulk voltage peak. Holes flood the channel leading to the source/drain voltage peak. The electrons that are trapped react much slower. They recombine with holes leading to a recombination current. The holes flow towards the source/drain, and the electrons towards the silicon gate oxide interface when the gate is biased towards accumulation again. The current that is measured over the complete pulse is the excess recombination current [56].

The charge pumping current (I_{CP}) is directly related to the interface trap density per energy (D_{it}), area (A), frequency (f), and the change in surface potential. The change in surface potential is related to the emission energy levels of electrons and holes (E_{em}, E_{hm}) (equation 3.12) [65]. The energy range is related to the rise time (t_{rise}), fall time (t_{fall}), pulse height (V_{Amp}), average capture cross section ($\sqrt{\sigma_e\sigma_h}$), and temperature (T).

$$
\begin{aligned}
I_{CP} &= eD_{it}fA(E_{hm} - E_{em}) \\
I_{CP} &= eD_{it}fA \cdot 2kTln\left(v_{th}n_i\sqrt{\sigma_e\sigma_h}\left|\frac{V_{th} - V_{fb}}{V_{Amp}}\right|\sqrt{t_{rise}t_{fall}}\right)\ for\ trapezoidal\ pulse
\end{aligned}
\tag{3.12}
$$

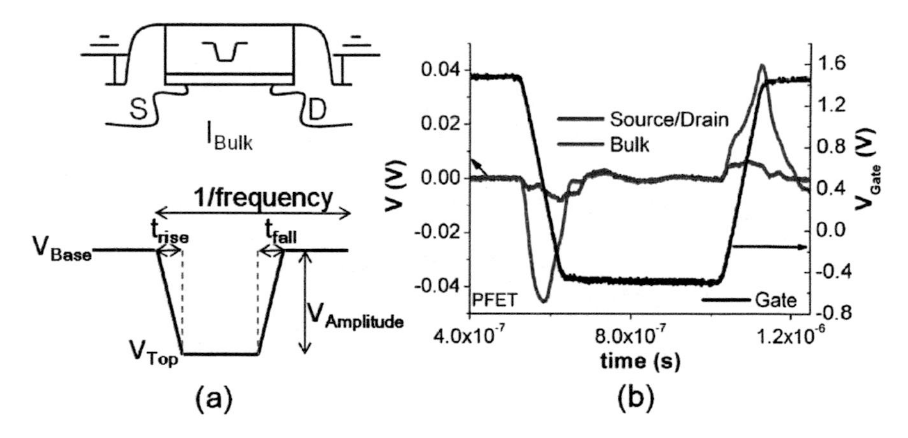

Figure 3.13: (a) Schematic of the measurement setup, and gate pulse characteristics are shown.
(b) Measured responses of source/drain-, and bulk voltage to the change in gate bias during CP. Measurements are done at a PFET (L_{Gate}=0.8μm, W_{Gate}=0.8μm) with a CET of 5.2nm at 25°C.

Different measurement methods are applied to analyze the CP current. For the so called base sweep, the pulse base voltage is increased, and the amplitude is kept constant. The charge pumping peak can be detected when the pulse base is below flatband voltage, and the pulse top is above threshold voltage (Fig. 3.14(a)) [66].

For the MultiFET, and overlap structures similar base sweep curves are obtained (see also Fig. A.5 of the appendix). Literature also indicates that short channel devices with inhomogeneous doping lead to higher errors in the CP measurements [68]. The devices of the overlap structures are used for the CP analysis.

Gate leakage current is detrimental for CP measurements of thin oxide devices (Fig. 3.14(a)). Several methods to compensate for gate leakage are discussed in this paragraph. The CP current is increasing with frequency (equation 3.12). Assuming that leakage current is frequency independent, the CP measurements can be corrected by subtracting a low frequency from a high frequency measurement [69]. However, the measured leakage current in this study is frequency dependent (Fig. 3.14). It is also suggested to extract the leakage current from the time dependent average of the gate to bulk leakage measurements [67]. The calculated leakage current exceeds the measured values (Fig. 3.14(a)).

The absolute value of the leakage current increases exponentially with voltage. The CP results in this study are corrected by fitting the leakage current with an exponential function as shown in figure 3.14(a).

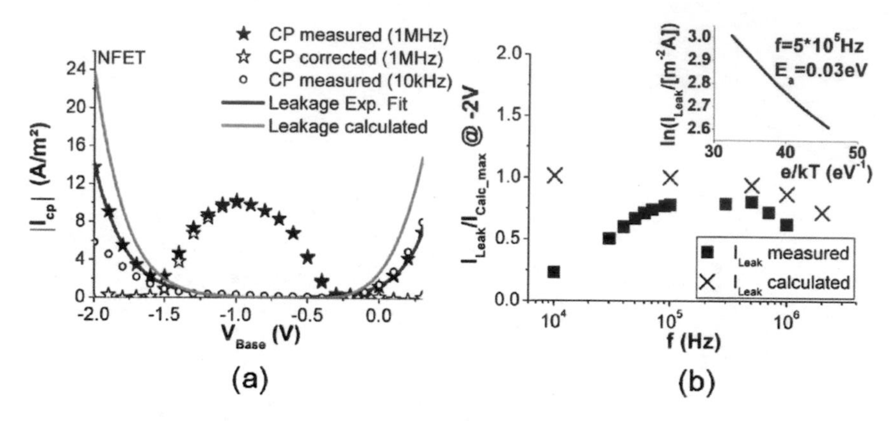

Figure 3.14: CP base sweep, and leakage interference measured at an NFET structure with a CET of 2nm at 25°C.
(a) Typical base sweep of a NFET with an amplitude of 1.5V, and pulse rise- and fall time of 100ns. The leakage current interference is measured at 10kHz, calculated [67], and fitted exponentially using the CP curve at 1MHz.
(b) Comparison of the calculated [67], and measured leakage current interference during CP analysis depending on the frequency under accumulation base bias. The temperature dependence of the measured leakage is shown in the inset.

In figure 3.14(b), the results of the leakage current calculation from the time dependent average of the gate to bulk current are compared to the measurements. The temperature dependence of the leakage current is investigated in the inset of figure 3.14(b). The activation energy of 0.03eV is similar to the activation energy of the static gate to bulk current at -2V of 0.07eV. Both leakages are caused by a band to band tunneling mechanism. For frequencies higher than 500kHz, the leakage current decreases as proposed in publication [67]. For a constant rise- and fall time of 100ns, the dwell time at the base- and the top voltage decreases with increasing frequency leading to a reduced leakage. As frequency increases, rise- and fall times are no longer negligible for the calculation of the overall gate current. For lower frequencies, the calculations estimate the leakage current to be constant (Fig. 3.14(b)). The measured current reduces far below the expected value for decreasing frequencies. This effect occurs only when the gate pulses are biased from accumulation to depletion.

The CP measurements in this study are done at 25°C, 100ns rise- and fall time, 1MHz, and with 50% duty cycle between V_{Top} and V_{Base}. The average interface charge (N_{it}) is calculated from the maximum CP current (equation 3.13) [70].

$$N_{it} = \frac{I_{CPmax}}{eAf} \tag{3.13}$$

A frequency dependent sweep with a constant rise- and fall time to pulse period ratio of 20% is performed to determine the average interface trap density, and the average of the capture cross section (equation 3.14) [71]. The measurements are done at the base voltage and amplitude of the maximum charge pumping current determined from base sweep.

$$\begin{aligned} t_{rise} &= t_{fall} = 0.2/f \\ D_{it} &= \frac{dI_{cp}}{dln(f/[Hz])}\frac{1}{2ekTfA} \\ \sqrt{\sigma_e\sigma_h} &= \frac{1}{v_{th}n_i}\left|\frac{V_{Amp}}{V_{th}-V_{fb}}\right|\frac{[f]_{I_{CP}(lnf)=0}}{0.2} \end{aligned} \tag{3.14}$$

The energy distribution of the interface traps near mid gap is determined by varying the rise- and fall time (equation 3.15) [66]. Measurements are performed at the base voltage and amplitude of the maximum CP current determined from the base sweep. The energy depth is calculated assuming an equal hole- and electron capture cross section (equation 3.14) [66].

$$\begin{aligned} E_t - E_i &= kT\left(v_{th}n_i\sqrt{\sigma_e\sigma_h}\left|\frac{V_{fb}-V_{th}}{V_{Amp}}\right|t_{fall}\right) \\ D_{it}(E_t - E_i) &= \frac{t_{fall}}{eAkTf}\frac{dI_{CP}}{dt_{fall}} \end{aligned} \tag{3.15}$$

To perform an amplitude sweep, the base voltage is kept constant below flat band voltage, and the top voltage is increased from below to above threshold bias. The CP current increases instantly when the threshold voltage is reached (Fig. 3.15). Side peaks in the derivative of the I_{CP} related to interface traps at the gate edge can be detected if the threshold voltage changes along the channel [72]. The threshold voltage changes due to silicon to gate oxide interface defects at the gate edge, or local dopant variations caused by the halo implants.

In reference [72], a method is developed to profile the interface defect density from the gate edge towards the channel center using an amplitude sweep. A source/drain voltage is applied when the gate is biased to accumulation. The electron density, and consequently the trapping probability during accumulation are significantly decreasing towards gate edge (Fig. 3.16). The extension of high electron density towards the gate edge during accumulation is dependent on the source/drain bias. A forward S/D diode voltage has to be applied to scan defects towards the gate edge. But, at high forward biases ($>|0.3|$V) at the source/drain CP measurements are impossible due to the forward diode current. So it is not possible to scan the overlap region with this method if the trap density at the gate edge is reasonably low. In this study, GIDL measurements are used to determine the silicon to gate dielectric interface trap density in the overlap region.

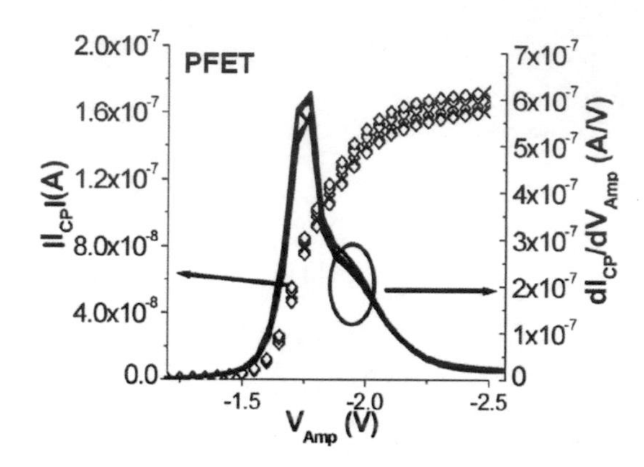

Figure 3.15: Amplitude sweep for a PFET with a CET of 5.2nm at 1MHz, and 250ns rise- and fall time. The derivation of I_{CP} reveals a side peak, due to V_{th} shifts along the channel.

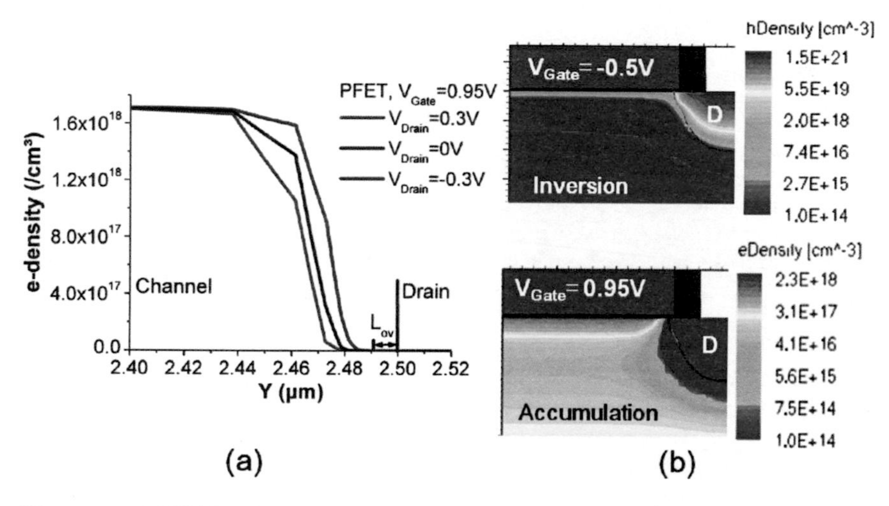

Figure 3.16: TCAD device simulation of PFET charge carrier density:
(a) PFET electron density under accumulation (V_{Gate}=0.95V) directly underneath the PFET gate at different source/drain biases is shown.
(b) Overview of the simulated PFET electron density under accumulation (V_{Gate}=0.95V), and hole density under inversion (V_{Gate}=-0.5V) underneath the PFET gate at V_{Source}=V_{Drain}=0V is presented.

3.4 Simulation

Simulations are used to separate the influences of changes in the electric field, and changes in defect density on the leakage current. The gate leakage is analyzed with the semi analytical BSIM4 model. The BSIM4 model is implemented in the ngspice software [55]. The spice simulations will be introduced in the first part of the section. A two dimensional model is needed to investigate the S/D leakage, SDE leakage, channel leakage, and GIDL current. The basics of the used TCAD process, and device software are explained in the second part of the section [54], [73].

3.4.1 Spice Simulation

The gate leakage is simulated using the program ngspice [55]. The BSIM4 model of the Berkeley university is implemented in ngspice [74]. A typical BSIM4 parameter set, called model card, of a PFET used in this study can be found in section A.5. Mainly the standard parameters that are recommended by BSIM4 for high-k transistors are used [74]. The parameters affecting the gate leakage are adjusted to fit the electrical data.

The BSIM4 model approximates the gate leakage semi analytically [74]. Direct tunneling, and Fowler Nordheim tunneling are taken into account. Different equations are used for the gate to channel, gate to bulk, and gate overlap current. Three fitting parameters in the BSIM4 model effect each of the gate leakages. With the ngspice simulations it is possible to fit the gate leakage of one transistor, and then compare the results with a similar device. Therefore, it is possible to investigate CET variabilities depending on the MOSFET geometry.

3.4.2 TCAD Simulation

TCAD Sentaurus process, and device simulations are performed to analyze the trap assisted S/D leakage, SDE leakage, channel leakage, and GIDL. The main production steps are included in the process simulation to obtain a realistic transistor structure. Standard TCAD models are used [73]. The gate oxide consists of SiO_2 with a thickness of the measured CET. The gate leakage current of the high-k devices is not modeled. The as implanted doses and energies of the dopants are used for the process simulations. Channeling, or transient enhanced diffusion of the dopants is not simulated to account for the germanium, and carbon implants. The thermal budget of the process flow is adjusted to give a good agreement between the simulated source/drain doping profiles, and the corresponding secondary ion mass spectroscopy measurements. A detailed description of a complete simulation process flow can be found in section A.6 of the appendix.

The DIBL shift of the transfer characteristic is calculated to check the validity of the process simulations. The TCAD device simulation includes the Philips mobility model with high field saturation, and transverse electric field dependence [54]. Band gap narrowing, and the effective intrinsic carrier density are also included. The hydrodynamic - and thermodynamic transport equation, and the Poisson equation are solved. A detailed description of a complete Sentaurus device command file, and model parameters are given in section A.7 of the appendix. The subthreshold currents at different drain biases are reproduced for the devices under test. The on-current fit depending on the mobility model is not optimized.

To analyze the trap assisted leakage current, a low constant mobility value is assumed, so no diffusion current occurs in the simulations. Since the devices operate at moderated voltages (0V to 3V), the avalanche multiplication model is not used. Band to band tunneling is calculated with the non local model over one path. The voltage dependence of the band to band tunneling current is fitted to the measured curves.

Different discrete defect regions with constant trap densities are placed at the silicon bulk, and at the dielectric to silicon interface. Trap assisted current generation after the Hurkx model is used in this simulations [38]. A further detailed discussion of the model parameters can be found in section A.8 of the appendix. The leakage current model is explained in more detail in the following chapters.

4 Impact of Implant Variations on PFET Leakage Current

Source/drain implants are adjusted in scaled transistors to reach shallow junctions, and reasonable threshold voltages. In the following chapter, the effects of the adjustments of the V_{th}- and halo implants, and the co-implantation of carbon on PFET leakage current are studied.

4.1 Process Flow

The peripheral DRAM transistors are prepared using Qimonda's 46nm process flow. The transistor is presented in figure 4.1. In this section, the most relevant process steps are described in detail (Fig. 4.2).

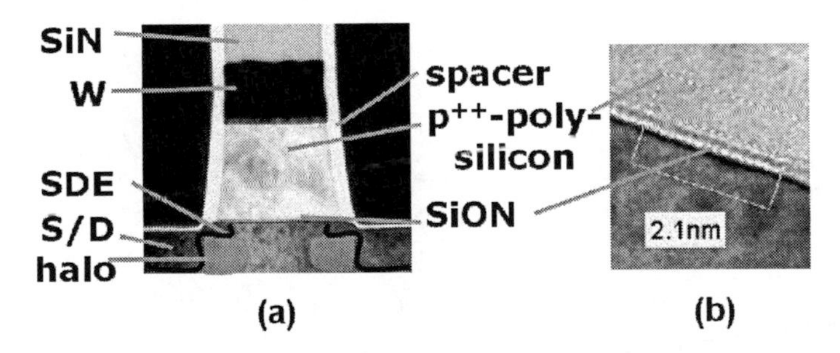

Figure 4.1: Transmission electron microscopy (TEM [75]) of the fabricated PFETs is presented.
(a) Overview of the transistor is presented.
(b) Magnification of the silicon oxynitride gate dielectric with a thickness of about 2nm is presented.

In the first step, a shallow trensh isolation is formed using silicon oxide. Subsequently, phosphorus and arsenic are implanted in the PFET well. A shallow arsenic implant for threshold voltage control follows. This implant will be called V_{th} implant in the following.

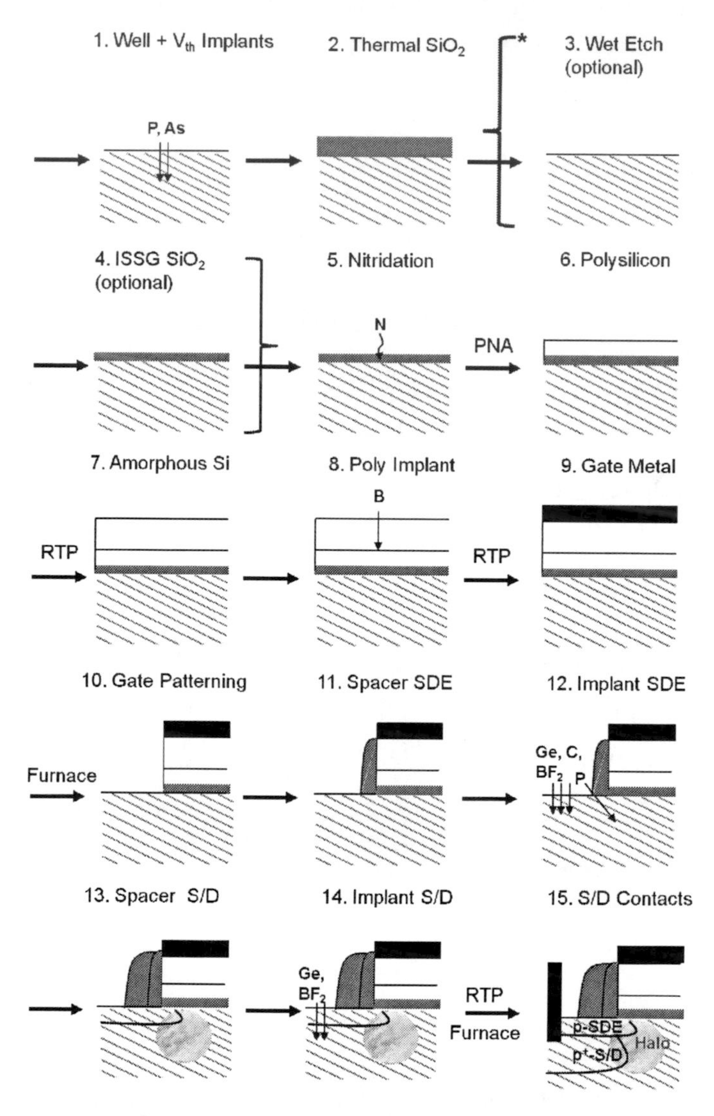

Figure 4.2: Most important process steps of the PFET fabrication for the devices with different junction implants are presented. A dual gate oxide process is used, growing either a thermal oxide, or an ISSG oxide (shaded: Si substrate, gray: (nitrided) SiO_x, white: polysilicon, amorphous Si, black: metal).

* Dual gate oxide process: Indicated process steps are used only for PFETs with thin gate dielectric.

The gate oxide is grown in the next step. Two different oxides are used for the peripheral DRAM transistors. First, a 5nm thick silicon oxide is grown by a thermal oxidation on all PFETs. For the transistors with reduced CET, the thermal oxide is removed by wet etch, and an in-situ steam generated (ISSG) silicon oxide of 2.5nm is grown. A decoupled plasma nitridation (DPN) is applied on both gate oxides to incorporate nitrogen in the gate oxide. The nitrogen is then transfered into silicon nitride bonds with a 1000°C post nitridation anneal (PNA) [76], forming a silicon oxynitride. The devices with a CET of 2.3nm, and 5.2nm will be called thin, and thick gate dielectric PFETs in this chapter.

The gate electrode is a polysilicon - tungsten nitride - tungsten stack. The polysilicon is deposited in two steps, due to the requirement to integrate the array together with the periphery devices. First, about 20nm polysilicon is deposited on top of the gate dielectric, followed by a rapid thermal anneal (RTA) step at 1000°C. The polysilicon surface is cleaned, and the second layer of 40nm amorphous Si is deposited. The polysilicon doping is achieved by a boron ion implantation. The boron is diffused towards the lower polysilicon layer, and electrically activated by an RTA step at 1000°C to obtain the p^+ polysilicon gate. A layer of tungsten nitride is sputtered on top of the polysilicon. The gate stack is completed by a sputtered tungsten layer, and a silicon nitride capping layer.

The gate stack is etched using a combination of dry and wet processes (Fig. 4.2 (10)). The thickness of the scattering oxide, remaining after the etch, can vary for different types of gate dielectric. A silicon oxide spacer of about 8nm is deposited from a tetra-ethyl-ortho-silicate (TEOS) precursor, and is etched in the next process step. The spacer separates the gate edge from the source/drain extension (SDE) implantation, and defines the overlap length.

Shallow SDE implants ensure a small overlap between the source/drain region, and the gate [77]. A germanium implant is used for amorphization of the silicon substrate. This avoids the channeling effect during the subsequent boron diflouride doping implantation [78]. The germanium amorphization introduces a high amount of interstitials. During the following processing, end of range (EOR) defects form at the amorphous/crystalline (a/c) interface [79]. The EOR defects serve as silicon interstitial reservoir, and the interstitials will be released during thermal treatment. These lead to two undesirable effects: Transient enhanced boron diffusion (TED) reduces the abruptness of the SDE profile [79], and boron interstitial clustering decreases the dopant activation [80]. Carbon is implanted to capture the released interstitials, and to suppress the growth of extended EOR defects [81]. The reduce TED efficiently, the carbon has to be positioned at a substitutional site in the lattice [82]. The incorporation of carbon in the silicon lattice is done during recrystallization of the amorphous region [83]. The increasing abruptness of the boron profiles in p^+n SDEs is studied extensively in literature [77, 80].

Halo implants are performed to increase doping concentration at the gate edge. Those implants are known to reduce short channel effects [84]. A tilt of 28° is used for the quadruple phosphorus implantation. The implantation also effects the threshold voltage in the short channel devices.

Table 4.1: Sample Description of the PFETs

Gate Dielectric	CET=2.3nm Split	CET=5.2nm Split
$V_{th}1$,C↑,Halo↔	V_{th} As dose ($1.5 \cdot 10^{12}$atm/cm^2)	V_{th} As dose ($1.1 \cdot 10^{12}$atm/cm^2)
$V_{th}2$,C↑,Halo↔ Standard Device	V_{th} As dose ($1.8 \cdot 10^{12}$atm/cm^2) C dose ($4.0 \cdot 10^{14}$atm/cm^2)	V_{th} As dose ($1.0 \cdot 10^{12}$atm/cm^2) C dose ($4.0 \cdot 10^{14}$atm/cm^2)
$V_{th}2$,C↓,Halo↔	C dose ($3.5 \cdot 10^{14}$atm/cm^2) Halo P dose ($2.3 \cdot 10^{13}$atm/cm^2)	C dose ($3.5 \cdot 10^{14}$atm/cm^2) Halo P dose ($2.3 \cdot 10^{13}$atm/cm^2)
$V_{th}2$,C↓,Halo↑	Halo P dose ($2.5 \cdot 10^{13}$atm/cm^2)	Halo P dose ($2.5 \cdot 10^{13}$atm/cm^2)
$V_{th}2$,C↓,Halo↓	Halo P dose ($2.1 \cdot 10^{13}$atm/cm^2)	Halo P dose ($2.1 \cdot 10^{13}$atm/cm^2)

Prior to the formation of the source/drain, an oxide spacer of about 33nm is deposited from a TEOS precursor, and etched. Germanium amorphization, and boron implantation are used to form the source/drain.

The dopants are activated using two RTA process steps at 1000°C. A final thermal process step is required in DRAM processing to reduce the defect density, and improve cell retention time. This DRAM anneal is a 30min furnace anneal at 800°C. The source/drain contacts are formed by a titanium/ titanium nitride liner, and tungsten filling.

This chapter focuses on the effect of the PFET implantation doses on the leakage currents of the transistors. In the first part of the chapter, the leakage current mechanisms are investigated in detail. The gate dielectric to silicon interface quality of the devices is analyzed. Two different carbon doses in the SDE, three different halo doses, and two different V_{th} doses are analyzed for both gate dielectrics in the second part of the chapter. A sample description is given in table 4.1.

Table 4.2: Implantation Energies and Doses for Standard PFET ($V_{th}2$,C↑,Halo↔)

Implant		[keV]	[atm/cm^2]
Well	P	340	$5.3 \cdot 10^{13}$
	As	30	$4 \cdot 10^{11}$
V_{th} (CET=2.3nm)	As	30	$1.8 \cdot 10^{12}$
V_{th} (CET=5.2nm)	As	30	$1 \cdot 10^{12}$
Poly	B	2.5	$6 \cdot 10^{15}$
SDE	Ge	20	$3 \cdot 10^{14}$
	BF_2	3	$4 \cdot 10^{14}$
	C	4	$4 \cdot 10^{14}$
Halo	P	40	$2.3 \cdot 10^{13}$
S/D	Ge	20	$3 \cdot 10^{14}$
	B	2.2	$2.5 \cdot 10^{15}$

The implantation energies, and doses for the standard PFET are given in table 4.2.

The total phosphorus halo dose of the quadruple implantation is given. The geometry parameters of the measured transistors can be found in table A.1.

4.2 TCAD Simulation of the Process

The main production steps are simulated with TCAD Sentaurus process. The temperature budget is adjusted to reproduce the boron doping profile measured by secondary ion mass spectroscopy (SIMS).

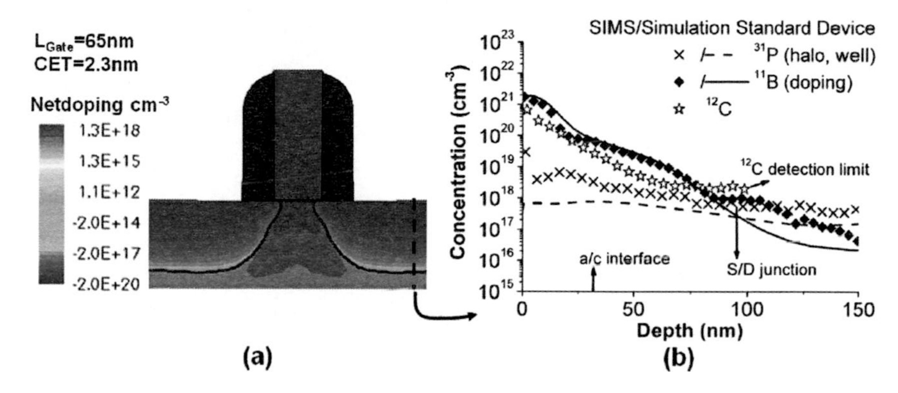

Figure 4.3: Simulated and measured doping profiles are shown.
(a) Transistor structure of a 65nm long PFET (CET=2.3nm) from process simulations
(b) SIMS profile of the S/D junction after processing [85] is presented. The process simulation gives a reasonable agreement with the measured doping gradients.

Figure 4.3 shows a comparison of the SIMS doping profiles of the source/drain junctions, and the simulated ones along the cut line. For the boron doping profile a reasonably good agreement is achieved. The measured halo doping concentration does not match the simulations. This is caused by the fact that the halo implantation dose is changed in the simulation to match the drain induced barrier lowering (DIBL) characteristics (Fig. 4.4). SIMS measurements are done on diode structure. In the diode structure, the quadruple phosphorus implant is not blocked at the gate stack. Therefore, a different halo doping is obtained by the SIMS measurements.

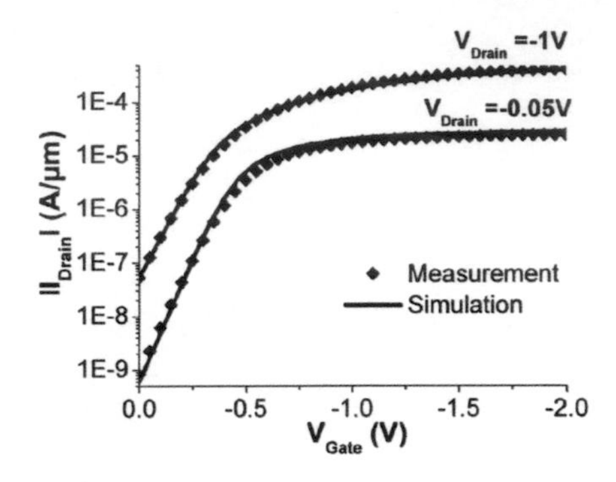

Figure 4.4: Subthreshold characteristic at different drain voltages is presented. A good agreement between measured, and simulated curves is reached for PFETs with a gate length of 65nm, and a CET of 2.3nm at 85°C.

4.3 Interface Traps

The interface trap concentration at the gate dielectric to silicon interface is measured by charge pumping (CP) technique. Figure 4.5 shows the measurement results of the average trap concentration per area (N_{it}). The PFET with thin gate dielectric has an increased interface defect density, compared to the PFET with thick gate dielectric.

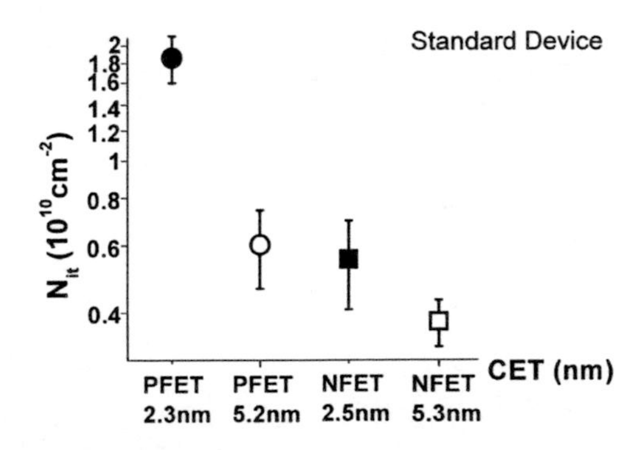

Figure 4.5: Gate dielectric to silicon interface trap density for PFET, and NFET with different gate dielectric thicknesses (NFET 1 table B.4). The interface trap density is determined at the maximum of the CP base sweep at 1MHz, 1.5V amplitude, and 100ns rise- and fall time.

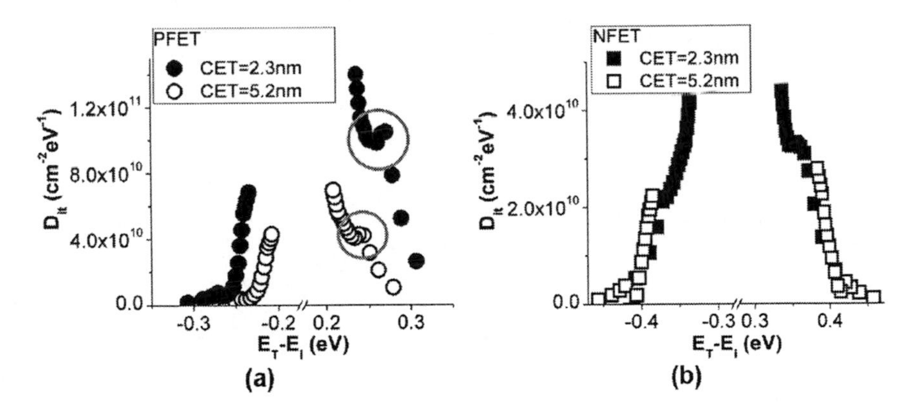

Figure 4.6: D_{it} determination by energy dependent CP measurements with an amplitude of ± 1.5V at 25°C for:
(a) A typical PFET ($V_{th}2$,C↓,Halo↔) with a capture cross sections of $2.3 \cdot 10^{-15}$cm^2, and $6.1 \cdot 10^{-15}$cm^2 estimated for the thin, and thick gate dielectric transistor, respectively.
(b) A typical NFET (NFET 1 from table B.4) with a capture cross sections of $2.6 \cdot 10^{-17}$cm^2, and $3.6 \cdot 10^{-18}$cm^2 estimated for the thin, and the thick gate dielectric transistor, respectively.

The NFET interface trap densities are shown in comparison. The NFET production is simlar to the PFET processing. Details on the NFET implantation steps can be found in table B.4 of the appendix. For the NFET, the defect concentration is increased by a factor of 1.5 with decreasing oxide thickness. For the PFET, the defect density is about a factor 3 higher for the thin gate dielectric transistors compared to devices with thick gate dielectric.

All NFET devices have a reduced interface defect concentration compared to the PFETs. One possible explanation is dopant diffusion from the polysilicon to the dielectric/polysilicon interface, or into the dielectric [86]. The nitridation reduces the boron diffusion more efficiently in the devices with a thicker gate dielectric [87]. The boron doping of the p^+ gate of the PFETs possibly leads to a higher defect concentration at the Si/SiO_2 interface compared to phosphorus doping in the polysilicon of the NFETs [88].

Energy dependent CP scans are preformed to get a better insight into the energetic levels of the defects. The measurements are done according to the method described in section 3.3. The change of defect concentration with the trap energy is presented in figure 4.6. The results agree well with published data [89]. The defect concentration at the conduction band edge is higher for the PFET with a lower CET. A defect band seems to occur at an energy range of 0.22eV to 0.28eV above mid gap. The energy dependent sweeps of the NFET samples show no increase of defect density towards the conduction band edge.

This indicates that the increase in interface defects in the PFETs with thin gate dielectric is due to the boron polysilicon doping.

One possibility discussed in literature is the formation of dangling bonds in a boron rich Si-SiO_2 interface [90]. In publication [91], dangling bonds, and the formation of boron oxygen complexes are discussed. The formation of boron rich positively charged defects from the polysilicon to gate dielectric interface is also discussed in literature [92].

The change in gate dielectric to silicon interface defect density is expected to effect the mobility. Figure 4.7 shows a decrease in mobility for the PFETs with thin gate dielectric at strong inversion, where charge carriers are scatter at surface roughnesses [62].

In conclusion, an increased silicon to gate dielectric interface trap density is measured for the PFET devices compared to the NFET devices. This increase is more evident for the transistors with thin gate dielectric. In the course of the investigation, a trap band at 0.22eV to 0.28eV above mid gap is found for the PFETs. The additional interface defects are possibly caused by the polysilicon boron doping. The interface trap density could be further reduced by a nitridation of polysilicon prior to annealing [93].

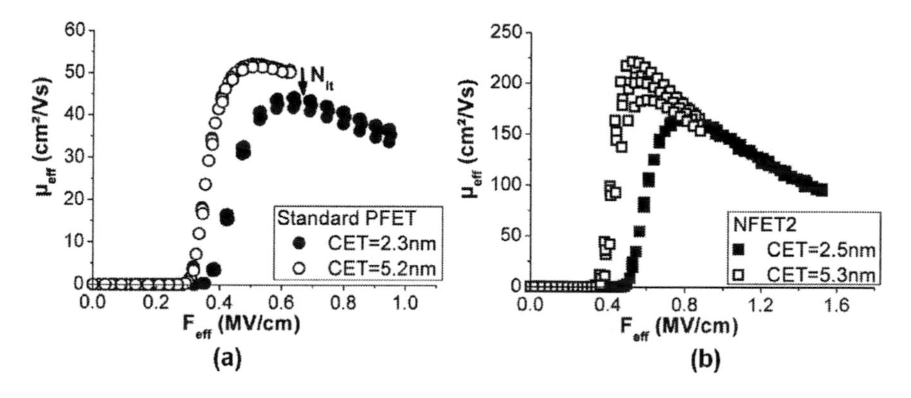

Figure 4.7: Mobility versus electric field measured at 85°C. Measurements are done at five transistors (L_{Gate}=5μm) each.
(a) PFET mobility depending on the gate dielectric thickness is presented.
(b) NFET (NFET 2 table B.4) mobility depending on the CET is shown.

4.4 Electrical Measurement and Simulation of Leakage Current

Off-current leakage occurs when the gate is biased to 0V. Different voltage conditions at the source/drain, and the bulk lead to different intrinsic leakage paths. Two main cases for peripheral DRAM transistors are presented in figure 4.8.

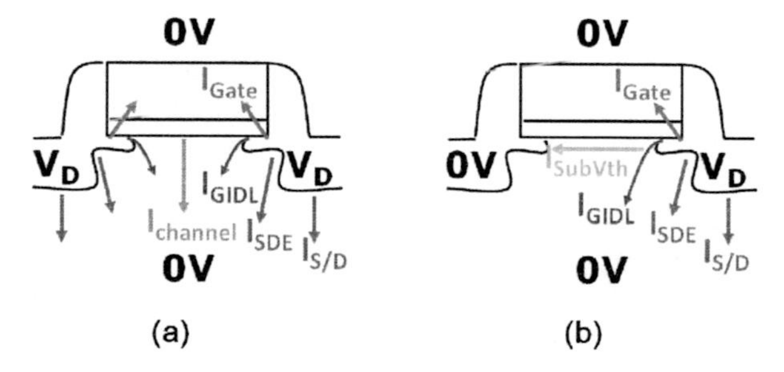

Figure 4.8: Schematic of the measured leakage currents occuring in off-state (V_{Gate}=0V), when:
(a) V_{Source}=V_{Drain}.
(b) V_{Source}=0V.

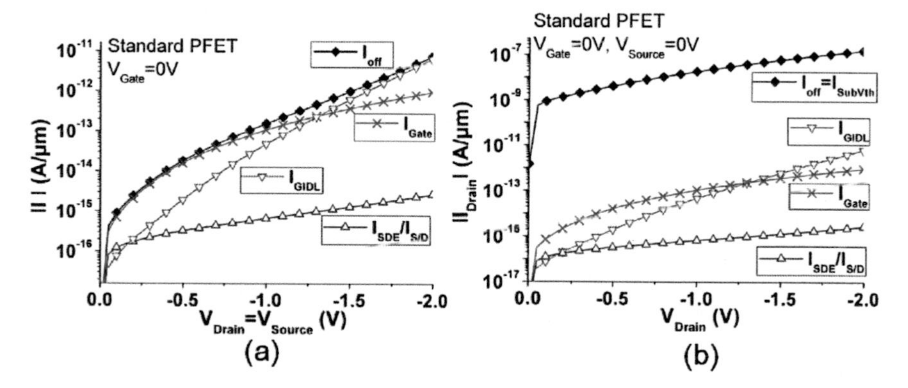

Figure 4.9: Typical leakage current voltage plot depending on the drain voltage for the 65nm long PFET with thin gate dielectric at 85°C, and V_{Gate}=0V. Different bias conditions are applied at the source:
(a) V_{Source}=V_{Drain}.
(b) V_{Source}=0V.

In the first case, source and drain are shorted when the voltage is ramped. In the second case, only drain potential is ramped while source is grounded. Figure 4.9 shows the different leakage current contributions to the total off leakage depending on the bias for the 65nm long PFET with thin gate dielectric. Equivalent measurements are done for the long channel PFETs with thin- and thick gate dielectric leading to similar results (section B.3 of the appendix). The dominant leakage current mechanisms of the total leakage depending on the bias are shown in figure 4.10. The most interesting case, equivalent to DRAM operation, is -1V and -3V at the drain for the thin and thick gate dielectric PFETs, respectively. Subthreshold-, gate overlap-, channel- and gate induced drain leakage current are the main contributions to the total peripheral transistor off-current.

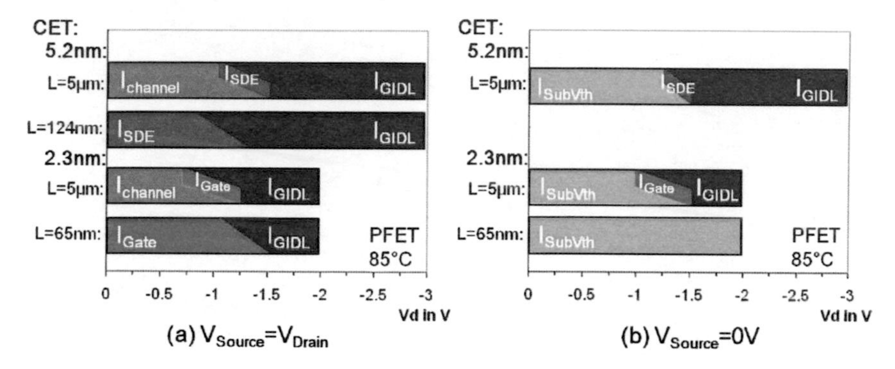

Figure 4.10: The bar graph shows the main leakage current contribution to the total off-current for short, and long channel PFETs depending on the drain voltage at 85°C. Different bias conditions are applied at the source:
(a) $V_{Source}=V_{Drain}$.
(b) $V_{Source}=0V$.

4.4.1 Source/Drain Leakage

The source/drain leakage ($I_{S/D}$) occurs at the junction of the vertical diffused source/drain (Fig. 4.11(a)). A diode test structure is used to measure S/D current (Fig. 4.11(b)). The advantage of the diode test structure is that the effective electric field can be determined from capacitance voltage measurements (section 3.2). Knowing the electric field, it is possible to analyze the leakage current mechanism.

The temperature (T) dependence of the S/D leakage current is given in Fig. 4.12(a). The activation energy (E_a) is calculated using equation (2.2). The calculated activation energy decreases with the voltage (Fig. 4.12(b)). Possible leakage current mechanisms are band to band tunneling (BTBT) [38], the Frenkel Poole effect (FP) [29], and a thermally assisted tunneling via defects (Hurkx) [38]. All mechanisms are discussed in section 2.3 in more detail.

Band to band tunneling has an activation energy below 0.15eV [32], and can therefore be excluded as leakage current mechanism. For the Frenkel Poole effect, the activation energy decreases linearly with the square root of the electric field [29]. The slope of the activation

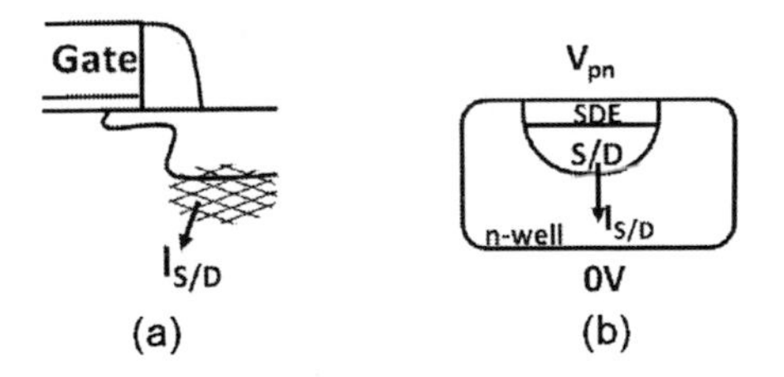

Figure 4.11: Measurement of the S/D leakage:
(a) S/D leakage current occurrence in the MOSFET is presented.
(b) Diode test structure for the measurement of the S/D leakage current is shown.

energy plotted against the square root of the electric field should be -22μeVm$^{1/2}$V$^{-1/2}$ [29] for a single charged defect. Figure 4.12(b) shows that the measured activation energy decreases rapidly with the voltage in contrast to the theoretical calculations for the Frenkel Poole effect. This decrease of activation energy with voltage can be explained using the Hurkx mechanism.

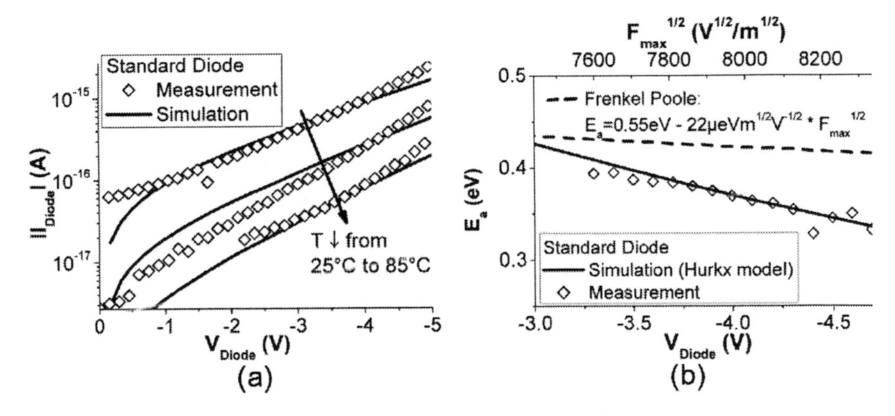

Figure 4.12: Measured, and simulated S/D leakage current of the p$^+$n standard diode:
(a) Measured, and simulated leakage currents in a T range from 25°C to 85°C are shown.
(b) Activation energy calculated from the temperature dependent measurements, and simulation. The theoretical calculation of the Frenkel Poole mechanism [29] results in a too low slope of the activation energy with the voltage.

The S/D diode is simulated by TCAD process simulation. The leakage current is calculated using the TCAD device software. Current due to the Hurkx mechanism, and direct band to band tunneling is simulated. All simulation parameters are listed in section B.4 of the appendix. The trap energy, and charge trapping efficiency, a parameter that gives the product of trap density and capture cross section, are extracted.

The simulated, and measured activation energies coincide (Fig. 4.12(b)). The simulated trap energy is at mid gap. The estimated trap density is $5.2 \cdot 10^{11} cm^{-3}$ for an assumed capture cross section of $1 \cdot 10^{-15} cm^2$ [94]. Also the voltage dependence of the simulated, and measured leakage currents match (Fig. 4.12(a)). At high diode voltages (-4.5V to -5V), the simulated leakage currents are lower than those measured. In the high voltage regime, an additional current increase due to avalanche multiplication is expected.

4.4.2 Source/Drain Extension Leakage

The source/drain extension leakage (I_{SDE}) occurs from the source/drain extension to the bulk (Fig. 4.13(a)). The Hurkx mechanism is used to simulate the trap assisted leakage in source/drain junction of the transistor [38]. Figure 4.14(a) presents the measured, and simulated leakage current from the S/D- and SDE junction of the PFET with a CET of 2.3nm, and a channel length of 65nm. The simulated leakage is about two decades lower as the measured current (lower curve of Fig. 4.14(a)), using a constant defect density of $5.2 \cdot 10^{11} cm^{-3}$ in the silicon bulk, as extracted from the S/D diode model. An additional defect band (N_I) is probably enhancing the SDE leakage current.

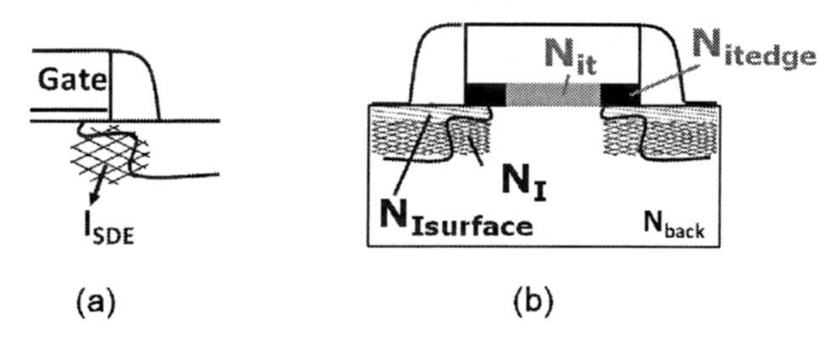

Figure 4.13: Schematic sketch of the traps responsible for generating SDE leakage:
(a) SDE leakage current occurrence in the MOSFET is presented.
(b) Trap distribution of the transistors assumed in the simulation. The SDE current is mainly generated in region N_I.

A trap rich region, which is labeled as N_I in figure 4.13(b), around the former amorphous/crystalline (a/c) interface at a depth 30nm is a possible cause of the increased leakage. It is known that end of range defects are formed at the a/c interface [95]. The SIMS results measured after SDE implantation prior to annealing reveal a local maximum of boron, and carbon at a depth of about 30nm (Fig. 4.15(a)). The maximum can be interpreted as a clustering of defects [96]. The clustering is dissolved during annealing (Fig. 4.15(b)). However, the results suggest that interstitial defects remain [97].

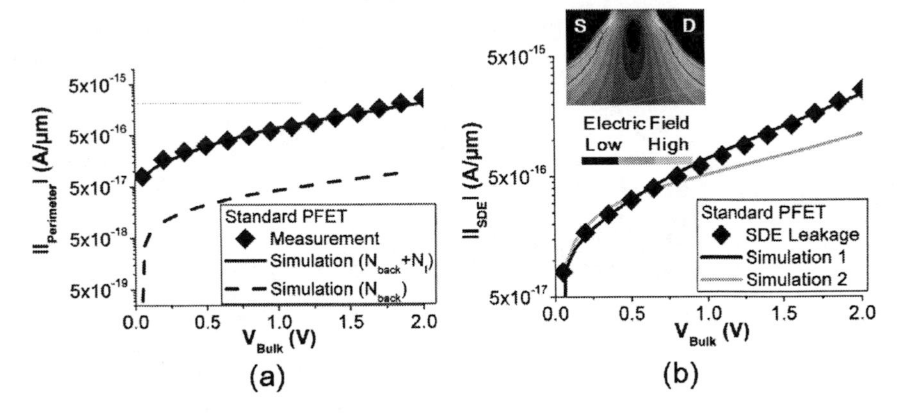

Figure 4.14: MOSFET perimeter leakage as measured at the PFET with the standard implants, a CET of 2.3nm, L_{Gate}=65nm, and W_{Gate}=5μm at 85°C.
(a) The measured MOSFET perimeter current is compared to simulations for different trap distributions (Fig. 4.13(b)).
(b) Simulated SDE current for different extensions of N_I into the channel is presented. In Simulation 1, N_I is placed accordingly to Fig. 4.13(b). In Simulation 2, N_I extends through the total channel, but the amount of defects in the depletion region is kept constant. The electric field variation over the channel is show in the inset of the graph.

The simulated, and measured SDE leakage current dependence on the voltage is matched if a trap rich region (N_I) is assumed (Fig. 4.16(a)). The activation energy, calculated from the model curves, exceeds the measured values (Fig. 4.16(b)). Again a mid gap trap is used. Model parameters are given in the section B.4 of the appendix. The placement of the defect region relative to the space charge region is as important as the defect density itself for the simulation.

The trap rich region (N_I) has to be between 20nm and 80nm from the silicon surface. If placed closer to the S/D junction, a too high diode current is obtained in the S/D simulations. Closer to surface, the defects can dissolve due to out diffusion of interstitial to the silicon surface [98]. If the defects are placed too far in the channel, the slope of the simulated leakage current with the voltage deviates from the measurements (Fig. 4.14(b)). This is caused by a different electric field in the channel (inset of Fig. 4.14(b)). The trap rich region occurs due to the implantation of the source and drain. So the defect density is expected to decrease towards the channel.

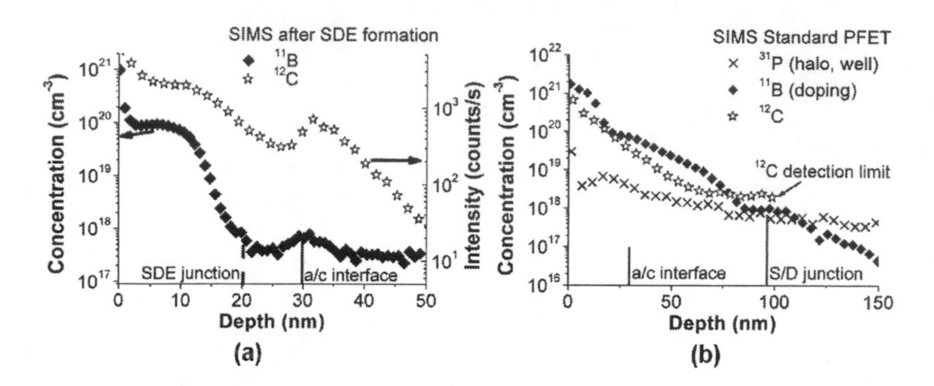

Figure 4.15: SIMS doping profiles [85] of the source/drain at different process steps: (a) SIMS profile after SDE formation prior to increased thermal treatment is presented. A local maximum of boron and carbon is observed at 30nm.
(b) SIMS profile at the end of processing is shown. The metallurgical S/D junction is between 90-100nm. The carbon concentration in the metallurgical junction is below the SIMS detection limit.

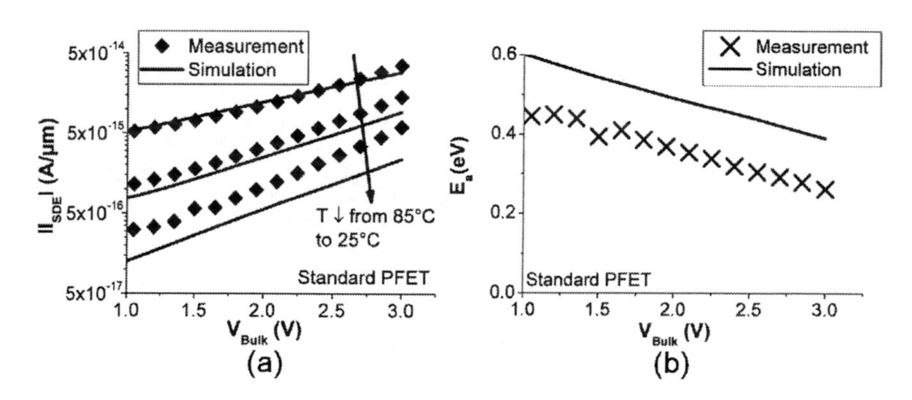

Figure 4.16: Measured, and simulated MOSFET perimeter leakage generated in the SDE depletion region. Measurements are done on a PFET with standard implants, and a CET of 5.2nm.
(a) Temperature dependence of the SDE leakage current is presented.
(b) Activation energy calculated from the temperature dependence of the SDE leakage is shown. The simulated exceeds the measured activation energy.

The simulated defect density, found to reproduce the experimental leakage of the thick gate dielectric PFET is $1.2 \cdot 10^{15}$cm^{-3} for an assumed capture cross section of $1 \cdot 10^{-15}$cm^{2} [94]. The simulated trap concentration of the thin p-channel transistors is $1.9 \cdot 10^{14}$cm^{-3} which is a factor 6.3 lower compared to the transistors with thick gate dielectric. This is unexpected because the same implant conditions are used for processing the source/drain region. The difference in trap concentration is too high to be interpreted only by local variations.

Section 4.1 suggests that thickness variations of the scattering oxide for the two dielectrics can occur prior to SDE implantation. The leakage currents corresponding to different etching depths are simulated (section B.5 of the appendix). Different scattering oxide thicknesses can not explain the increase in leakage current by a factor of 6 in the simulation.

4.4.3 Generation Leakage from Channel Region to Bulk

The so called channel leakage ($I_{channel}$) is a generation current which occurs in the depletion region at the gate area (Fig. 4.17(a)). The PFET channel current is generated at interface defects, or bulk defects close to the interface (section 2.3). It is simulated with the Hurkx model assuming traps at silicon to gate dielectric interface (Fig. 4.17(b)). Defects with a uniform energy distribution of ΔE=0.05eV around mid gap are used. In the model, the fit parameter of the effective mass is increased in case of the interface traps compared to the bulk silicon defects. The fitted effective mass is consistent with literature values for leakage currents generated at interface traps [51, 99]. Simulation parameters are given in section B.4 of the appendix.

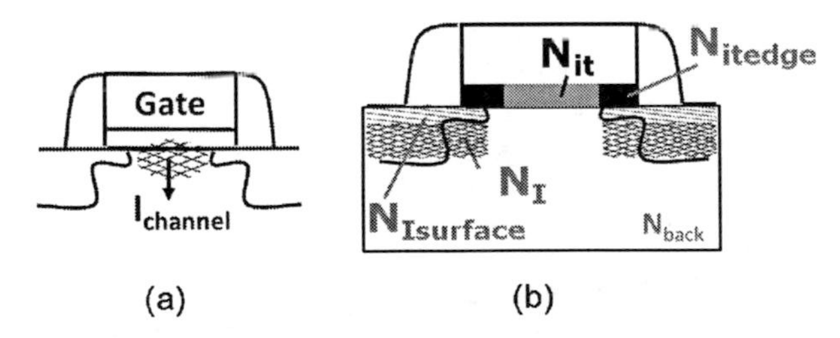

Figure 4.17: Sketch of the traps responsible for generating the channel leakage:
(a) Channel leakage current flow from the depletion region beneath the gate.
(b) Trap distribution of the transistor assumed in the simulation. The channel leakage is mainly caused by N_{it}.

A significant influence of the channel leakage on the overall current is found only in long devices (L_{Gate} >500nm). The current is due to defects in the middle of the channel. $I_{channel}$ has a very small dependence on the substrate bias due to the low electric field of

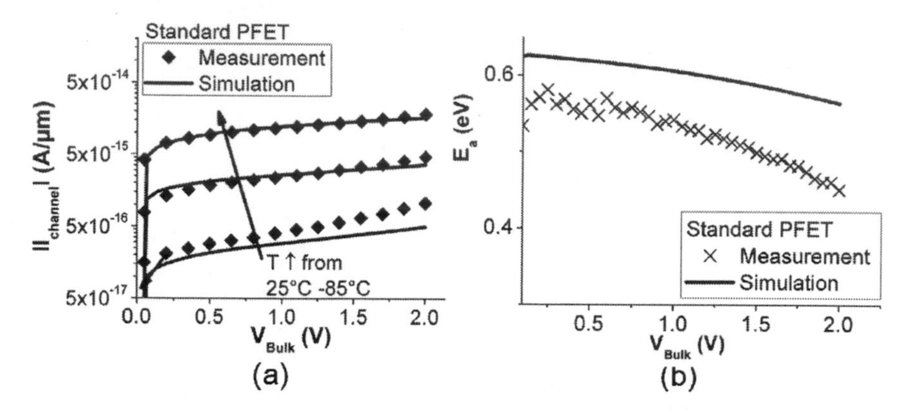

Figure 4.18: Measured, and simulated channel leakage for a PFET with standard implants, a CET of 2.3nm, and a gate length of 5μm.
(a) Temperature dependence of the channel current is presented.
(b) Activation energy calculated from the temperature dependence of $I_{channel}$.

around 0.4MV/cm at a voltage of +1V for a 5μm long thin gate dielectric PFET in the channel middle. The measured, and simulated temperature dependence of the channel leakage exhibit the same trend (Fig. 4.18). An offset in activation energy is observed. The offset is caused by an overestimation of the gate dielectric to silicon interface trap concentration, and/or an underestimation of bulk channel defect density. Further investigation of the defect gradient from the channel to the bulk silicon would require additional test structures.

The interface trap densities that are fitted in the simulations are compared to the interface trap densities measured by charge pumping. Therefore, a capture cross section of $5{\cdot}10^{-17}$ cm^2 is assumed in the simulations. The interface trap density from the simulation is $3.7{\cdot}10^{10}$eV^{-1}cm^{-2} for the thin-, and $0.7{\cdot}10^{10}$eV^{-1}cm^{-2} for the thick gate dielectric. The charge pumping measurements give a defect density of $3.02{\pm}0.70{\cdot}10^{10}$eV^{-1}cm^{-2} in the thin-, and $0.98{\pm}0.32{\cdot}10^{10}$eV^{-1}cm^{-2} in PFETs in the thick gate dielectric. The ratio in interface trap density between the two gate dielectrics is 4.3 in the charge pumping measurements, and 5.3 for the simulations. The ratio is in the same range both in simulations, and measurements.

4.4.4 Gate Induced Drain Leakage

The gate induced drain leakage (GIDL) is especially sensitive to defects at the gate dielectric to silicon interface of the gate edge (Fig. 4.19). The density of defects at the gate

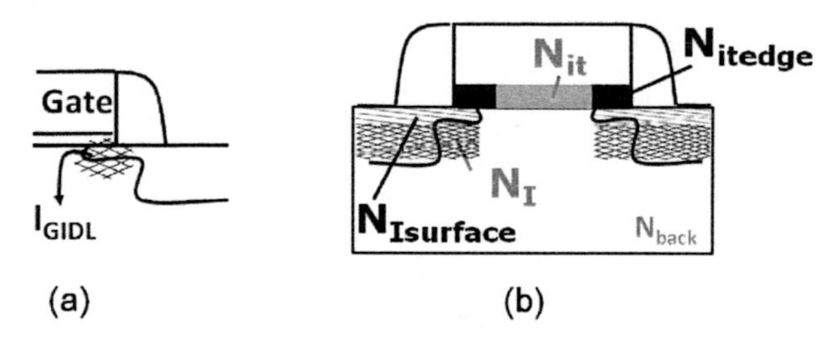

Figure 4.19: Sketch of the traps responsible for generating GIDL current:
(a) GIDL current flow from the S/D overlap region to the bulk.
(b) Trap distribution of the transistor assumed in the simulations. GIDL is mainly caused by N_{itedge} and $N_{Isurface}$.

edge is expected to be higher than in the channel area, and effects the device reliability [72].

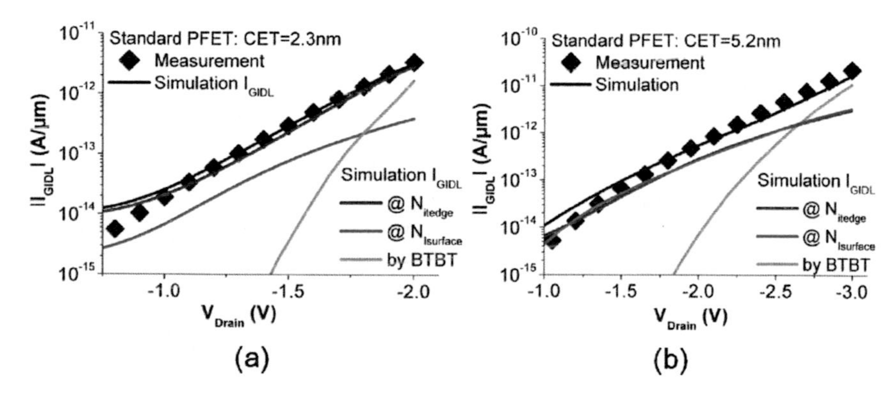

Figure 4.20: GIDL current measurement, and simulation for PFETs with standard implants at 85°C. The contribution of N_{itedge}, and $N_{Isurface}$ (Fig. 4.19) to the overall leakage current is presented for a:
(a) PFET with a CET of 2.3nm.
(b) Devices with a CET of 5.2nm.

The GIDL current occurs at a deep depletion that is formed in the overlap region of the S/D extension underneath the gate (section 2.3). There are three components influencing the GIDL (Fig. 2.10). Band to band tunneling is almost not temperature dependent with an activation energy below 0.15eV [32]. Trap assisted tunneling where charge carriers tunnel through the remaining barrier after thermal activation (Hurkx model) [51]. Trap

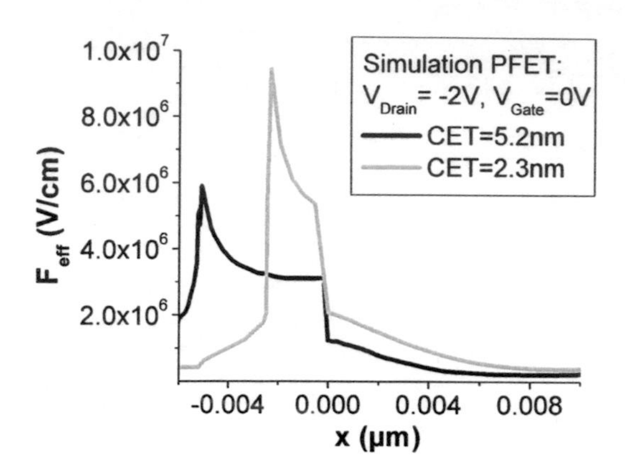

Figure 4.21: Simulated vertical electric field for PFET with thin-, and thick gate dielectric at the gate edge at -2V.

assisted tunneling occurs when the bulk traps are close to the surface ($N_{Isurface}$) as well as in case of interface traps at the gate edge (N_{itedge}, Fig. 4.19(b)).

The GIDL current is simulated for a 5μm long devices using the Hurkx model [38]. The simulated, and measured leakage current increase with voltage is in good agreement (Fig. 4.20). A one level mid gap trap is used for the bulk silicon trap simulations. For the interface defects a uniform energy distribution of ΔE=0.05eV around mid gap is assumed. The estimated defect concentrations from the model are given in table 4.3. The simulation parameters are listed in section B.4 of the appendix.

In the simulation, both the lateral leakage from the interface traps (N_{itedge}), and the horizontal leakage bulk traps ($N_{Isurface}$) are considered. For the PFET with a CET of 5.2nm, both currents contribute equally to the GIDL (Fig. 4.20(b)). For the PFET with a CET of 2.3nm, mainly the interface trap dependent current contributes to the overall GIDL (Fig. 4.20(a)). For a similar voltage range, the electric field at the interface is much higher for the thin gate dielectric samples leading to a higher current from the interface traps (Fig. 4.21).

The measured, and simulated temperature dependences are in good agreement for the PFETs with thin gate dielectric (Fig. 4.22(a)). For the current caused by a thermally assisted tunneling mechanism via defects, the activation energy decreases with the voltage down to 0.15eV. Once the activation energy reduces below 0.15eV, the band to band tunneling mechanism dominates the leakage current [32]. For transistor with thick gate dielectric, the simulated activation energy exceeds the measured at lower voltages (Fig. 4.22(b)). One possible explanation is that the N_{itedge} are placed too far in the channel, leading to an additional leakage from the channel that exceeds the GIDL at low voltages.

Table 4.3: Concentration of Defects Leading to GIDL Current for Standard PFETs

Trap region	gate dielectric	Trap Concentration	Capture Cross Section
$N^{*}_{Isurface}$	CET=5.2nm	$5.0 \cdot 10^{13} cm^{-3}$	$1 \cdot 10^{-15} cm^{2}$
	CET=2.3nm	$8.0 \cdot 10^{12} cm^{-3}$	$1 \cdot 10^{-15} cm^{2}$
D^{**}_{itedge}	CET=5.2nm	$2.6 \cdot 10^{12} cm^{-2} eV^{-1}$	$5 \cdot 10^{-17} cm^{2}$
	CET=2.3nm	$4 \cdot 10^{11} cm^{-2} eV^{-1}$	$5 \cdot 10^{-17} cm^{2}$

*N - trap density in the silicon (cm^{-3})
**D - trap density per energy ($cm^{-2}eV^{-1}$)

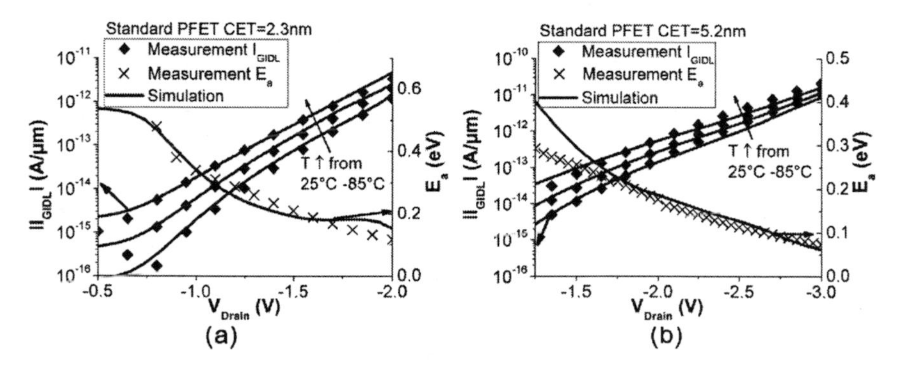

Figure 4.22: Temperature dependence, and activation energy of the GIDL current for the PFET with standard implants is presented for a:
(a) PFET with a CET of 2.3nm.
(b) Devices with a CET of 5.2nm.

4.4.5 Verification of the Defect Model: Gated Diode Measurement

The gated diode measurement (section 3.1) is used to study the generation of charge carriers in different regions of a MOSFET [28]. The measurements are performed on the PFET with thick gate dielectric to validate assumptions that are made in the TCAD model used in the simulations. The resulting measured, and simulated currents in the bulk depending on the gate voltage are compared (Fig. 4.23).

When the gate is biased from inversion up to around -0.7V, the leakage current is mainly due to the depletion region around the source/drain junction. The measured current is within the noise level for the 5μm long transistor structures. The PFET is in depletion for gate voltages between -0.7V and -0.05V. When the gate is biased to depletion, the gated diode current arises from the silicon to gate dielectric interface defects, and traps very close to the interface. The simulated current exceeds the measured by a factor of 1.5 due to an overestimation of the interface trap density in the model. When the gate is biased to accumulation, the current is increasing with the voltage mainly due to GIDL current coming from interface traps, and bulk defects close to the silicon surface at the gate edge. The gated diode measurements, and simulations agree. The assumptions of the TCAD simulation model describe the change of bulk current with gate voltage accurately.

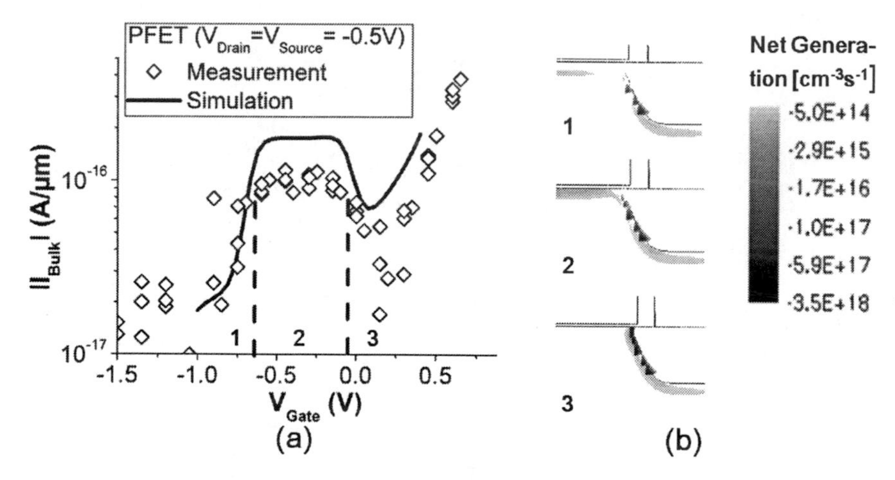

Figure 4.23: Gated diode measurement at five dies, and simulation for PFETs (V_{th}2,C↓,Halo↔) with a CET of 5.2nm, and a gate length of 5μm at 25°C.
(a) Gated diode leakage at the bulk is presented. The gated diode current is within the measurement resultion when the gate is biased to inversion.
(b) The net generation rate for the inversion (1), depletion (2), and accumulation (3) case is shown.

4.4.6 Gate Leakage

The gate leakage (I_{Gate}) is unaffected by the implant conditions of the devices. It has a high influence on the overall off-current for the PFETs with thin gate dielectric. No gate leakage above the measurement noise level is detected for the transistors with thick gate dielectric.

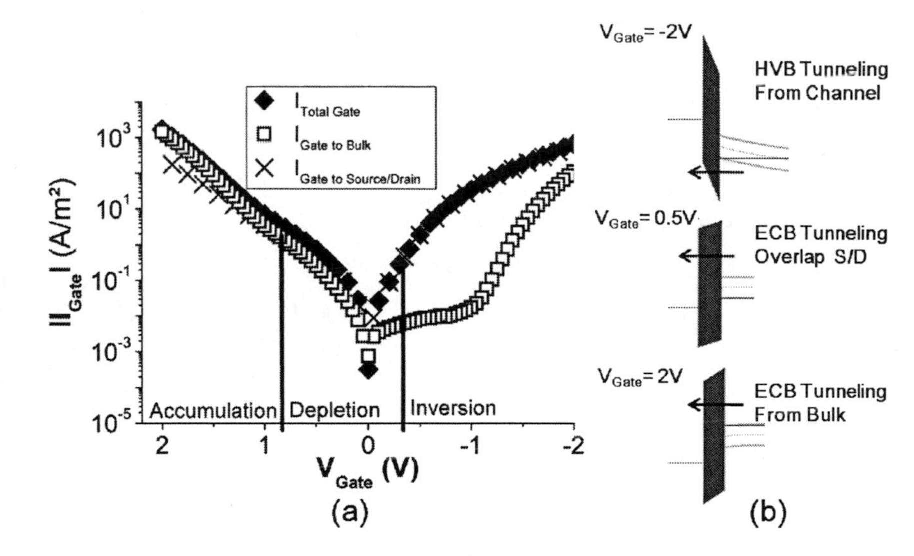

Figure 4.24: Gate leakage of a 65nm long PFET (V_{th}2,C↓,Halo↑, CET=2.3nm) at 85°C is presented.
(a) Contribution of gate to source/drain, and gate to bulk leakage on overall current depending on the voltage are shown.
(b) Sketch of the leakage flow due to hole valence band tunneling (HVB) at V_{Gate}= -2V, and electron conduction band tunneling (ECB) at V_{Gate}=2V is presented. The source/drain overlap current dominates the depletion regime (V_{Gate}= 0.5V) [35].

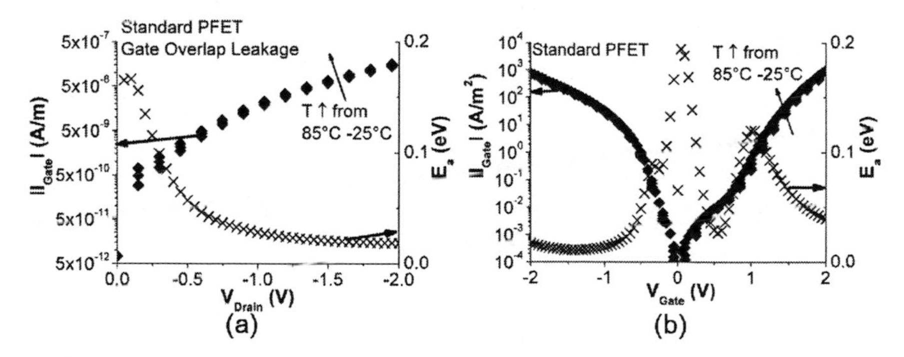

Figure 4.25: Temperature dependence, and activation energy of the gate leakage for the 5μm long PFET with standard implants, and a CET of 2.3nm.
(a) Gate to source/drain overlap leakage is presented.
(b) Overall gate leakage in a temperature range from 25°C to 85°C is shown.

Figure 4.24 presents the dependence of the gate leakage on the voltage for the 65nm long PFET. When the gate is biased, the current can be divided in three regions (Fig. 2.2(a)). In accumulation, an area dependent current flow, caused by electron conduction band tunneling (ECB) towards the bulk, is observed. In the depletion regime, an ECB leakage over the source/drain overlap occurs. In inversion, an area dependent current flow, caused by hole valance band tunneling (HVB) towards the channel, arises [35]. The 5μm long PFETs are investigated in the same manner. The results of this investigation can be found in section B.7 of the appendix.

The activation energy calculated from the temperature dependent current voltage curves of the gate current is below 0.15eV (Fig. 4.25). Therefore, a band to band tunneling causes the leakage current [32]. Mainly two conduction mechanisms are observed: Fowler Nordheim tunneling through a triangular barrier, and direct tunneling through a trapezoid barrier (section 2.1) [26]. A Fowler Nordheim HVB current is found under strong inversion (Fig. 4.24(b)). In strong accumulation a direct tunneling mechanism is observed (Fig. 4.24(b)).

An overlap current flows under off conditions. The overlap leakage is enhanced for the short channel PFETs under test (Fig. 4.26). In comparison the NFET test samples don't show that kind of behavior.

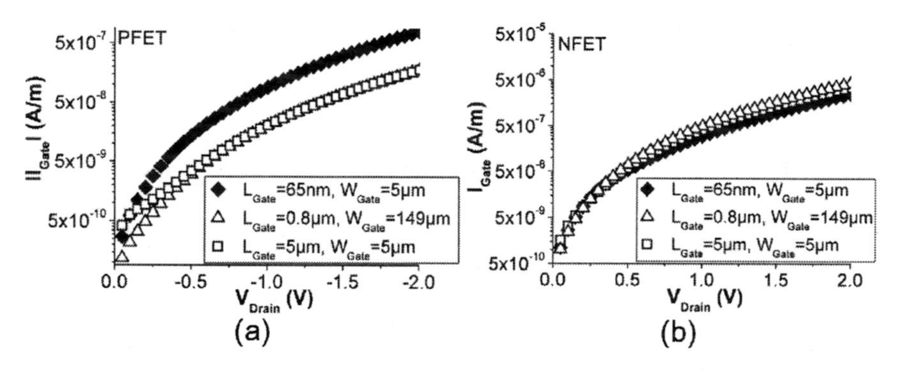

Figure 4.26: Channel length dependence of the gate to source/drain overlap leakage current for transistors with a CET of 2.3nm at 85°C.
(a) PFETs (V_{th}2,C↓,Halo↑) show an increased overlap leakage for short channel devices.
(b) For the NFETs (NFET2 table B.4) the overlap leakage scales with the gate width.

4.4.7 Subthreshold Leakage

The subthreshold leakage (I_{SubVth}) is a current between source, and drain when the gate bias is below V_{th}. The subthreshold current is mainly dominated by the diffusion current [3]. It is enhanced by the drain induced barrier lowering effect (DIBL). The DIBL effect accounts for the threshold voltage reduction depending on the drain bias (equation 2.5) [37]. The DIBL parameter (λ) is dependent on the type of gate dielectric, the gate length, and the doping.

The subthreshold current is calculated with using equation (4.1) [100]. The DIBL parameter is used as a free fitting parameter. It determines the slope of the subthreshold current with the drain voltage. The calculated, and measured currents are in good agreement (Fig. 4.27). The fitted DIBL parameter agrees well with the one extracted from the threshold voltage change of the transfer curves. Both the subthreshold current, and the DIBL effect have a strong dependence on the V_{th}- and halo implants.

$$I_{SubVth} = \mu_{eff} C_{ox}(m-1)\frac{W}{L}\left(\frac{kT}{e}\right)^2 \cdot exp\left[\frac{e(V_{Gate} - V_{th} + \lambda \cdot V_{Drain})}{mkT}\right] \cdot \left(1 - exp\left[\frac{-eV_{Drain}}{kT}\right]\right) \tag{4.1}$$

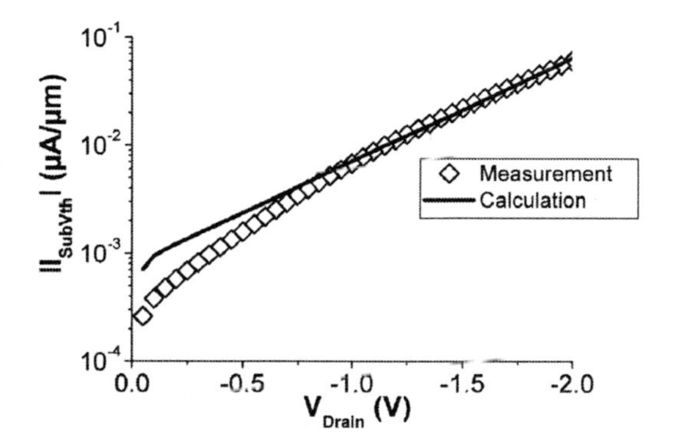

Figure 4.27: Measured, and calculated subthreshold leakage for the 65nm long PFET (V_{th}2,C↓,Halo↑) with a CET of 2.3nm at 85°C. A good agreement between measured DIBL parameter (λ=0.11), and the fitted value (λ=0.115) from the calculation is reached. Threshold voltage and body factor, used for the calculation, are determined from the transfer curves at V_{Drain}= -0.05V.

4.4.8 Conclusion

In this section, the various leakage currents under off conditions (Fig. 4.8) in PFETs with different implants have been investigated. The gate-, and subthreshold leakage are analyzed with standard methods. The gate leakage is independent of the implantation conditions.

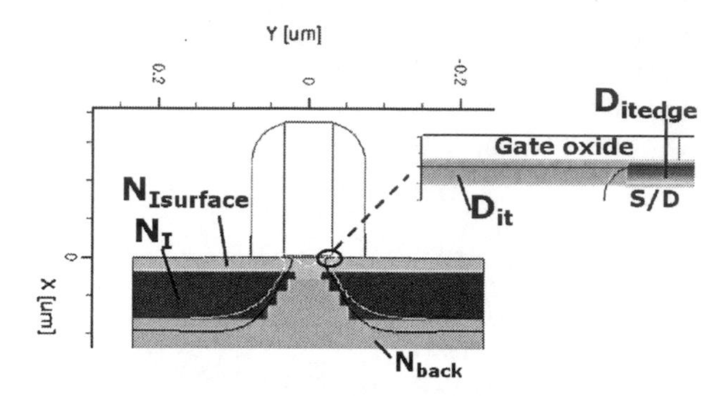

Figure 4.28: Placement of the continuoes defect bands in the simulations presented for a PFET with a CET of 2.3nm, and L_{Gate}=65nm.

A TCAD Sentaurus simulation model with different defect regions is developed (Fig. 4.28) to explain source/drain-, source/drain extension-, channel-, and gate induced drain leakage (GIDL) current. It is based on a generation current enhanced by trap assisted tunneling (Hurkx [38]). Additionally, direct band to band tunneling is included in the model. The model explains the different leakage current paths, their temperature-, and voltage dependence. The Frenkel Poole effect, avalanche multiplication, and the saturation current are not needed to reproduce the leakage current characteristics.

The trap distribution in the model is discrete, using trap bands with constant defect density (Fig. 4.28). The accurate extraction of the real values of the trap density is ambitious because of two effects. First, the capture cross section is not known. Second, the simulation reacts sensitively to the fit value of the trap density, and the placement of the discrete bands. The advantage of the model is that the leakage currents, and defect densities of two similar devices can be compared. In addition, a rough comparison of the trap concentration in different defect regions of one transistor is possible. Table 4.4 gives an overview of the extracted defect concentrations for the standard PFETs. The capture cross section is assumed to be constant for the bulk silicon defects, and different for the silicon to dielectric interface defects in this model.

The model is based on mid gap traps in the bulk silicon. The trap density is enhanced around the source/drain extension (N_I), probably due to interstitial defects from the germanium amorphization implantation. Small stable interstitial clusters with an energy around mid gap among others have been reported [97, 101, 102]. The defect clusters are too small to be observed by transmission electron microscopy [101]. The amount of interstitial clusters after annealing drops significantly below the former amorphous/crystalline interface towards the source/drain junction [97].

For the gate dielectric to silicon, and the spacer oxide to silicon interface traps a small uniform energy distribution of 0.05eV around mid gap is assumed. The model confirms the expected trap density increase at the gate edge (N_{itedge}) [72], comparing the channel-, and the gate induced drain leakage (table 4.4).

Table 4.4: Concentration of Defects for Standard PFETs

Trap Region	Gate Dielectric	Trap Concentration	Capture Cross Section	Trap Energy
N^*_{back}	Both	$5.2 \cdot 10^{11}$cm^{-3}	$1 \cdot 10^{-15}$cm^2	0.55eV
N_I	CET=5.2nm	$1.2 \cdot 10^{15}$cm^{-3}	$1 \cdot 10^{-15}$cm^2	0.55eV
	CET=2.3nm	$1.9 \cdot 10^{14}$cm^{-3}	$1 \cdot 10^{-15}$cm^2	0.55eV
$N_{Isurface}$	CET=5.2nm	$5.0 \cdot 10^{13}$cm^{-3}	$1 \cdot 10^{-15}$cm^2	0.55eV
	CET=2.3nm	$8.0 \cdot 10^{12}$cm^{-3}	$1 \cdot 10^{-15}$cm^2	0.55eV
D^{**}_{it}	CET=5.2nm	$7.0 \cdot 10^{9}$cm^{-2}eV^{-1}	$5 \cdot 10^{-17}$cm^2	0.55eV$\pm$0.05eV
	CET=2.3nm	$3.7 \cdot 10^{10}$cm^{-2}eV^{-1}	$5 \cdot 10^{-17}$cm^2	0.55eV$\pm$0.05eV
D_{itedge}	CET=5.2nm	$2.6 \cdot 10^{12}$cm^{-2}eV^{-1}	$5 \cdot 10^{-17}$cm^2	0.55eV$\pm$0.05eV
	CET=2.3nm	$4.0 \cdot 10^{11}$cm^{-2}eV^{-1}	$5 \cdot 10^{-17}$cm^2	0.55eV$\pm$0.05eV

*N - trap density in the silicon (cm^{-3})
**D - trap density per energy (cm^{-2}eV^{-1})

4.5 Carbon Implantation into Junction Extension

Carbon is implanted into the SDE to reduce the boron diffusion leading to a more abrupt doping profile for better short channel control (section 4.1). Since the carbon remains it also effects the boron diffusion during the S/D formation (see also Fig. B.8). A change of doping profile at the metallurgical junctions leads to changes in the electric field, and so to different leakage currents. Additionally, the leakage currents are possibly enhanced by carbon related defects.

4.5.1 Description of Experiment

The aim of this experiment is to investigate the effect of carbon co-implantation on the defect distribution in the PFETs. Carbon rich silicon potentially reduces the formation of extended EOR defects [81], but is also known to form smaller defect clusters which increase leakage currents [103]. Samples with two different carbon implantation doses, $4 \cdot 10^{14}$atm/cm^2 (High C Dose), and $3.5 \cdot 10^{14}$atm/cm^2 (Low C Dose), are investigated. A small difference in the carbon doses is chosen to keep the change in the electric field minor. So the variations in leakage current due to carbon related defects can be investigated.

All the other implantation conditions are summarized in table B.1 of the appendix. Figure 4.15 shows the SIMS profile after SDE formation, and at the end of the process. A local maximum of boron, and carbon is observed at 30nm after SDE formation. An a/c interface depth of 28nm is calculated from the germanium implantation conditions with the computer program SRIM [104], adding up ion range and ion straggle. The local maximum of boron, and carbon is probably an interstitial clustering (EOR defects) prior to increased thermal treatment [79]. After the complete thermal budget, no local maximum at 30nm is observed anymore (Fig. 4.15). The clustering is probably healed out during the annealing steps [105]. The metallurgical S/D junction depth is between 90-100nm. The carbon concentration in the metallurgical junction is below detection limit of the SIMS measurements.

4.5.2 Electrical Characterization of Transistor Performance

The effects of the change in carbon dose by a factor of 1.15 on the basic electrical parameters are investigated in this section. The most important parameters for the short channel PFETs with a CET of 2.3nm are summarized in table 4.5. A comparison of the long channel devices with thin-, and thick gate dielectric can be found in the tables B.8, B.9 of the appendix.

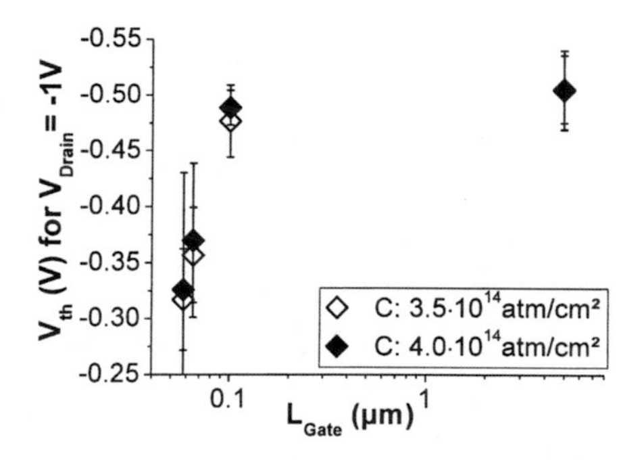

Figure 4.29: Saturation V_{th} versus channel length for transistors with a CET of 2.3nm at 85°C. The PFETs with higher carbon dose show in average a slightly better roll off behavior.

The transistors with increased carbon dose show on average a slightly better short channel control, and improved roll off behavior (Fig. 4.29). However this improvement is within the statistical deviation of the parameters. The same result can be found in table 4.5. From literature it is expected that the boron diffusion in the S/D region is altered with the carbon dose. Judging by the results for the junction capacitance (C_{Diode}), and overlap length (L_{ov}) no difference in dopant profile is found. The junction capacitance would react sensitively to modifications in the lateral dopant profile of the S/D (Fig. B.9 of the appendix). The overlap length is a measure of the horizontal boron diffusion in the SDE.

Table 4.5: Basic Characteristics of the PFETs with Different C Implants (L_{Gate}=65nm, CET=2.3nm)

	C: 4.0·10^{14}atm/cm^2	C: 3.5·10^{14}atm/cm^2
CET (nm)	2.32±0.02	2.32±0.01
L_{ov} (nm)	6.3±0.2	6.5±0.8
C_{Diode} (fF/μm^2) V_{Diode}=-1V (CV*)	1.28±0.02	1.27±0.02
V_{th} (V) V_{Drain}=-0.05V (CV**)	-0.40±0.05	-0.37±0.08
V_{th} (V) V_{Drain}=-0.05V (IV)	-0.49±0.02	-0.48±0.02
V_{th} (V) V_{Drain}=-1V (IV)	-0.37±0.07	-0.36±0.04
V_{fb} (V) (CV**)	0.9±0.05	0.9±0.05
I_{on} (μA/μm)	185.4±47.8	192.7±25.7
S_{Vth} (mV/dec) V_{Drain}=-0.05V	117.2±8.1	118.7±11.2
λ (IV***)	0.124±0.056	0.13±0.03

All values are given for measurements on 5 dies (wafer center).
* Capacitance voltage measurements are done on a diode structure.
** V_{fb} and V_{th} are taken from the point of maximum, and minimum slope of the CV.
*** DIBL parameter is determined from the V_{th} shift of $I_{Drain}V_{Gate}$ as described in section 2.2.

So, the author assumes that the modification of the boron doping profile is below 1nm. A dose of 3.5·10^{14}atm/cm^2 seems to be sufficient to effectively capture the interstitial defects, and reduce dopant diffusion.

4.5.3 Interface Traps

Figure 4.30 summarizes the results of the CP measurements for the PFETs with different carbon implants. The average gate dielectric to silicon interface trap density per energy (D_{it}) is determined by frequency dependent measurements, as described in section 3.3.

Transistors with a gate length of 0.8μm are used for the analysis. The average D_{it} is not effected by the carbon implantation dose. Possible carbon related interface defects at the gate edge [106] can not be detected by measuring PFETs with an area to perimeter ratio of 0.4. The CP current is directly proportional to measured area (equation 3.12). The effective contribution from the perimeter is about 12.5% of the total CP current as

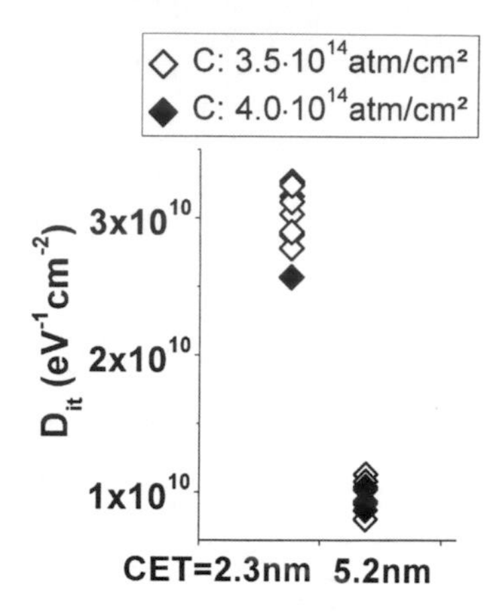

Figure 4.30: Dependence of the channel interface trap density on the carbon dose for the two different oxide types. No carbon related increase in trap density is found by frequency dependent CP measurements.

derived from equation (4.2), assuming a carbon diffusion of 50nm into channel. The CP current contribution from the gate edge is within the measurement variation.

$$I_{CPedge} = I_{CPtotal} \cdot \frac{A_{edge}}{A_{total}} = I_{CPtotal} \cdot \frac{12500\mu m \cdot 0.05\mu m}{5000\mu m^2} = 0.125 \cdot I_{CPtotal} \quad (4.2)$$
$$(Thin\ PFET:\ L_{Gate} = 0.8\mu m\ \rightarrow Table\ A.1\ of\ the\ Appendix)$$

A variable amplitude sweep is preformed to determine the lateral distribution of the interface traps along the channel. Figure 4.31 compares the derivation of the CP current with the pulse amplitude for the PFETs with different gate dielectrics. Side peaks, related to interface traps at the gate edge, can be detected if the threshold voltage changes along the channel, and a high amount of gate edge interface defects is present (section 3.3 [72]). In Fig. 4.31(b), a side peak is observed for the PFET with a CET of 5.2nm. For the PFET with a CET of 2.3nm no such peak occurs.

Fig. 4.32 shows the side peaks in dependence of the carbon dose. The variation of the side peak is within the statistic variability of the measurement. With the lateral CP technique, the channel can be investigated up to the metallurgical SDE junction (section 3.3). The measurement is restricted up to the overlap length of 7.3nm into the channel. The change of the CP current derivation due to the carbon dose reduction is small if traps 7.3nm away from the gate edge are scanned.

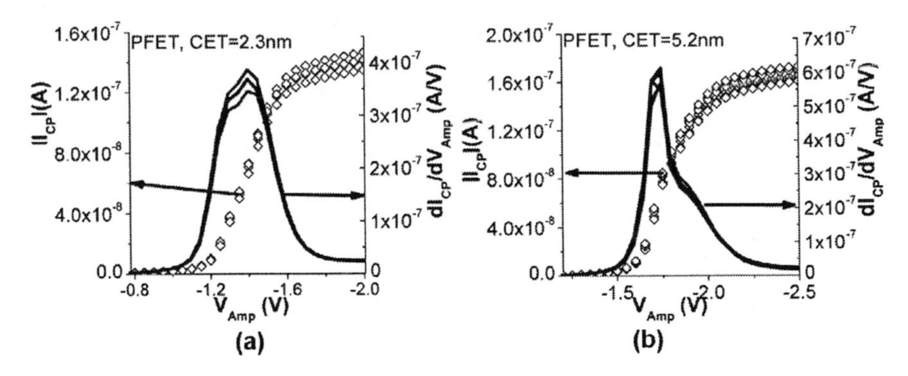

Figure 4.31: Increase of CP current with increasing voltage amplitude, and its derivation for PFETs with a low carbon dose ($3.5{\cdot}10^{14}$atm/cm^2). Pulses were applied with a frequency of $1{\cdot}10^6$Hz, and 250ns rise- and fall time at 25°C.
(a) Results for PFETs with a CET of 2.3nm are shown.
(b) Results for PFETs with a CET of 5.2nm are presented. The side peak in the derivation is related to the interface traps at the gate edge.

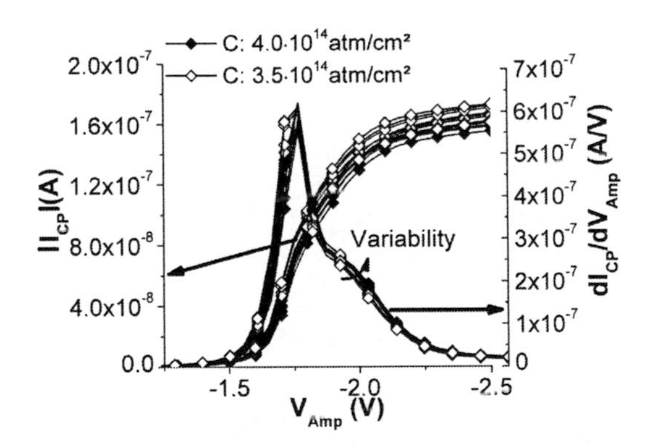

Figure 4.32: Increase of CP current with increasing voltage amplitude, and its derivation for PFETs with a CET of 5.2nm. Pulses were applied with a frequency of $1{\cdot}10^6$Hz, and 250ns rise- and fall time at 25°C. Within the measurement variation, no difference in the curves is observed due to the change in carbon implantation dose.

4.5.4 Carbon Dose Dependent Leakage Current

Carbon is implanted during source/drain extension formation. The changes of S/D-, SDE-, and GIDL current depending on a carbon dose modification by a factor of 1.15 are investigated in this section.

Source/Drain Leakage

The S/D junction leakage is not effected by the carbon implant (Fig. 4.33). The pn-junction is placed approximately 90nm to 100nm below the silicon surface (Fig. 4.15). The carbon concentration in the junction is below the SIMS detection limit. Even by the electrical measurements, which are more sensitive to active defects than SIMS, no carbon related defects are observed in the space charge region of the S/D junction [107].

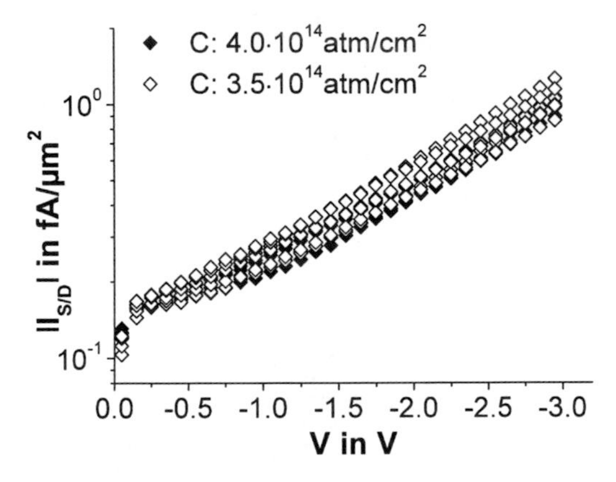

Figure 4.33: Carbon dose dependence of the p^+n-junction leakage at 85°C. Within the measurement variation of the samples, no difference in leakage current is observed.

Source/Drain Extension Leakage

Fig. 4.34 presents the SDE leakage for two different gate dielectrics. The SDE leakage current from the trap rich region increases with the carbon implantation dose. One possible explanation is that the incorporation of carbon stabilizes the defects [108]. So more defects remain after annealing. Also carbon atoms can form additional defects, and these could explain the increase in leakage current [109]. A carbon dose dependence of the S/D leakage is not measured, which indicates that the carbon related trap concentration rapidly decreases towards the S/D junction.

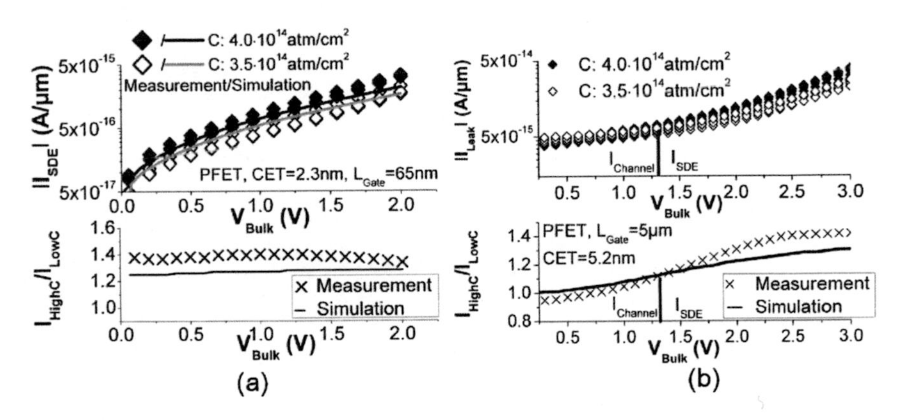

Figure 4.34: Measured SDE leakage current at 85°C in dependence on the carbon dose dose. Measured leakage current is shown for five dies each (upper part of the graph). The average measured, and simulated leakage current ratios are in agreement (lower part of the graph).
(a) Results of thin gate dielectric PFETs are shown.
(b) Results of thick gate dielectric PFETs are presented.

The simulated, and measured leakage current ratios, between PFETs with high carbon -, and PFETs with low carbon implantation dose, are compared in figure 4.34(b). The average measured leakage current ratio is 1.4 in the SDE regime for the thick gate dielectric transistors. In the simulation model, a leakage current increase due to a defect density increase is assumed [110]. In the model, the leakage current increase equals the defect density increase in the trap rich regions N_I and $N_{Isurface}$ (Fig. 4.13). The simulated leakage current ratio is 1.3 for the thick gate dielectric PFETs. It is in good agreement with the measurements. The simulated, and measured leakage current ratio for the PFETs thin gate dielectric is 1.3 (Fig. 4.34(a)).

A change of carbon implantation dose by a factor of 1.15 leads to a defect density increase by a factor of 1.3 in the simulations. This finding indicates that every additional implanted carbon leads to two extra electrical active defects. The carbon implantation doses are $3.5 \cdot 10^{14}$atm/cm^2, and $4 \cdot 10^{14}$atm/cm^2 for both oxide. The change in source/drain extension leakage current is independent of the oxide type. It is concluded that there are carbon related defects in the source/drain extension depletion region.

Gate Induced Drain Leakage

An increase of the GIDL current with increased carbon dose is found experimentally (Fig. 4.35, and 4.36). The error of the GIDL current measurement is within the size of the symbol (see also Fig. B.6 of the appendix). Trap assisted tunneling (Hurkx) is the main GIDL mechanism. An increase in trap density directly leads to a constant increase in GIDL current. With increasing electric field, the contribution of direct BTBT, which is a defect independent mechanism, to the total GIDL is rising.

The GIDL current ratio between PFETs with high carbon -, and PFETs with low carbon implantation dose is constant in the trap assisted regime. A decrease of the GIDL ratio is observed for the voltage range where band to band tunneling is the main leakage mechanism. For the simulated curves, the ratio decreases much faster at higher voltages compared to the measured results. This finding indicates that the band to band tunneling component is overestimated in the simulations. The trap assisted GIDL increases by a factor of 1.3 analysing samples with thin gate dielectric, and a factor 1.8 investigating devices with thick gate dielectric.

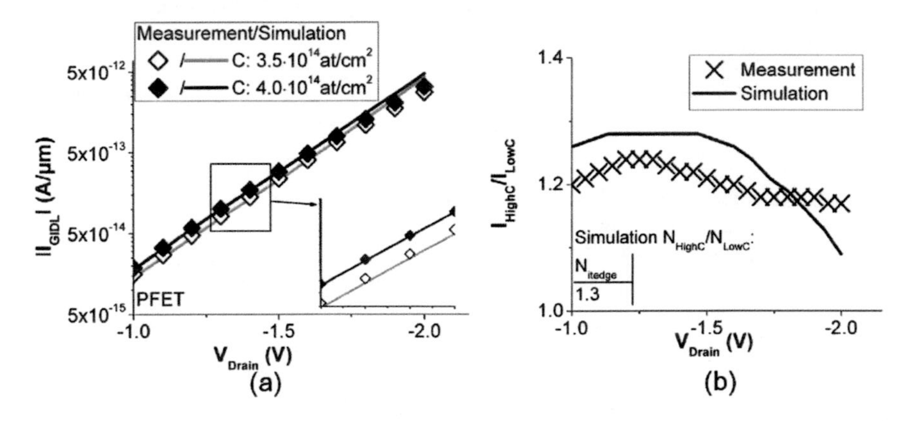

Figure 4.35: Measured GIDL current at 85°C in dependence on the carbon dose dose for the PFETs with a CET of 2.3nm.
(a) Comparision of the measured, and simulated GIDL current. The measurement error is within the symbol size (see also Fig. B.6).
(b) The average measured, and simulated change of leakage current ratio with the drain bias are shown.

The carbon implant leads to a reduced transient enhanced diffusion. From the simulation results, the N_{itedge} assisted GIDL current of the thin gate dielectric devices is constant or slightly decrease if the overlap length is reduced by 2nm (Fig. 4.37(a)). The decrease is caused by the lower boron doping concentration of the source/drain in the deep depletion region. The decrease in active doping leads to a smaller electric field at the gate edge interface (Fig. 4.37(b)). The simulation results are contrary to the measurement results. The increased GIDL is caused by an increase in defect density.

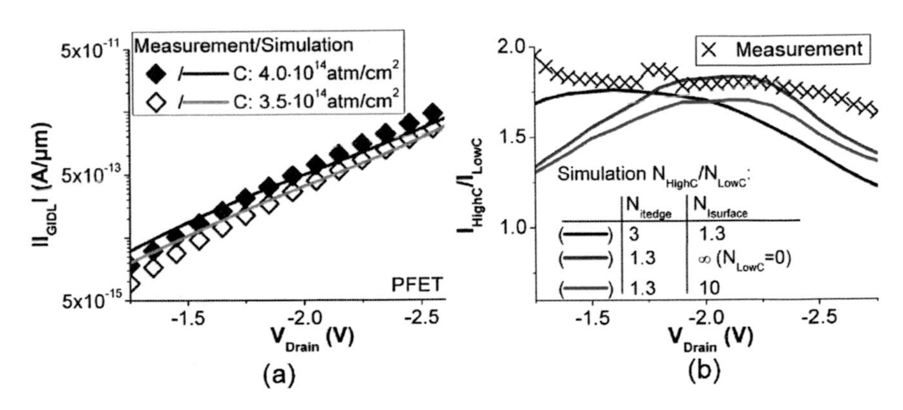

Figure 4.36: Measured GIDL current at 85°C in dependence on the carbon dose for the PFETs with a CET of 5.2nm.
(a) Comparision of the measured, and simulated GIDL current. The measurement error is within the symbol size (see also Fig. B.6).
(b) The average measured, and simulated leakage current ratios are in good agreement below -2.25V.

The model assumes a change in silicon to spacer and gate dielectric interface defect density at the gate edge (N_{itedge}), as well as an increase in bulk silicon traps ($N_{Isurface}$) due to the carbon implant (Fig. 4.19) [110]. A carbon induced increase in interface trap density has been suspected before [111]. For the transistor with thin gate dielectric, the GIDL current is mainly due to a Hurkx generation via the N_{itedge} (Fig. 4.20(a)). In this case, the ratio in trap assisted GIDL between the high -, and low carbon dose PFETs corresponds to the interface trap density ratio. The change in N_{itedge} with the carbon implantation dose is implemented in the model. The measured, and simulated GIDL current ratio of 1.3 are in good agreement (Fig. 4.35).

For the thick gate dielectric PFETs, both bulk ($N_{Isurface}$), and interface traps (N_{itedge}) have to be taken into account (Fig. 4.20(b)). A bulk trap ratio due to the carbon implantation dose of 1.3 is determined by the SDE leakage current measurements. The measured GIDL current ratio is 1.8. An interface trap density increase by a factor of 3 has to be assumed to fit the measured GIDL current ratio in that case (Fig. 4.36). However, it is also possible that the change of $N_{Isurface}$ with carbon implantation dose is higher than estimated from the SDE experiments. The SIMS profile shows that the carbon concentration is highest at the silicon/spacer interface (Fig. 4.15(b)). The GIDL current ratio is reproduced assuming an interface trap density increase between 1.3 and 3 for different $N_{Isurface}$ ratios (Fig. 4.36).

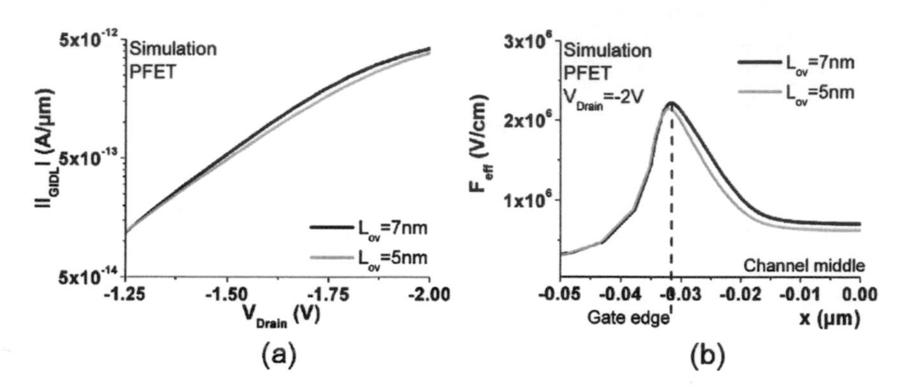

Figure 4.37: Simulated change in GIDL characteristics with overlap length for a 65nm long PFET with a CET of 2.3nm at 85°C.
(a) No change, or a small decrease in interface trap assisted GIDL with smaller overlap length is found.
(b) The variation of electric field from the gate edge towards the channel middle at the interface depending on the overlap length are shown.

4.5.5 Conclusion

This part of the chapter focuses on the effect of carbon co-implantation in the source/drain extension on electrical characteristics of PFETs. Carbon implantation doses of $3.5{\cdot}10^{14}$ atm/cm^2, and $4{\cdot}10^{14}$atm/cm^2 are used for the investigation. The small difference in carbon dose allows to analyze the leakage current behavior without a drastic change in diffusion behavior.

Table 4.6: Measured Leakage Current Ratios for PFETs with different Carbon Doses

CET=2.3nm		*Leakage Current Ratio at V_{Drain}=-1V and V_{Gate}=0V				
Implant	Dose (atm/cm^2)	GIDL	SDE	Channel	Gate	Subthreshold
C	$4.0{\cdot}10^{14}/3.5{\cdot}10^{14}$=1.15	1.3	1.3	1	1	1 (L_{Gate}=65nm) 1/1 (L_{Gate}=5μm)

CET=5.2nm		*Leakage Current Ratio at V_{Drain}=-3V and V_{Gate}=0V				
Implant	Dose (atm/cm^2)	GIDL	SDE	Channel**	Gate	Subthreshold
C	$4.0{\cdot}10^{14}/3.5{\cdot}10^{14}$=1.15	1.6	1.4	1	-	1/1.1 (L_{Gate}=5μm)

*Average measured leakage current ration is given.
** Channel current ratio measured at -0.75V for thick dielectric PFETs.

The change in the basic electrical characteristics with the increase in carbon concentration is within the statistical variations of the measurements (table 4.5). No modification in vertical diffusion are found experimentally. The effect of the carbon dose on the overlap length (horizontal diffusion) is also within the measurement statistics, so a change in L_{ovl} below 1nm is assumed.

The gate induced drain leakage, and source drain extension leakage are increased with higher carbon implantation dose. The channel leakage, source drain current, subthreshold current, and gate leakage are unchanged (table 4.6).

The increase in leakage current with higher carbon dose is interpreted as, an increase in interface defect density at the gate edge (N_{itedge}), and an increase in defect density around the SDE (N_I, $N_{Isurface}$) in this study. The simulation results of the changes in defect density are summarized in table 4.7. It could be that carbon stabilizes the defects. That would explain the increase in interface trap density due to carbon implantation which was also found by other authors [111]. It is possible that carbon gathers interstitials and forms small defect complexes in the SDE region [108]. Another explanation for the increased defect density is the formation of carbon interstitial, and phosphorus substitutional pairs [109]. Trap energies around mid gap have been found for these kind of defects [109].

No carbon related defects are found in the source/drain junction, at 90nm-100nm below the silicon surface, by electrical measurements. The defects caused by the carbon implantation do not diffuse that far in the silicon bulk.

Table 4.7: Simulated Trap Ratios for the PFETs with Different Carbon Implants

Defect Region	Ratio	Placement
N_{back} N_{HighC}/N_{LowC}	1	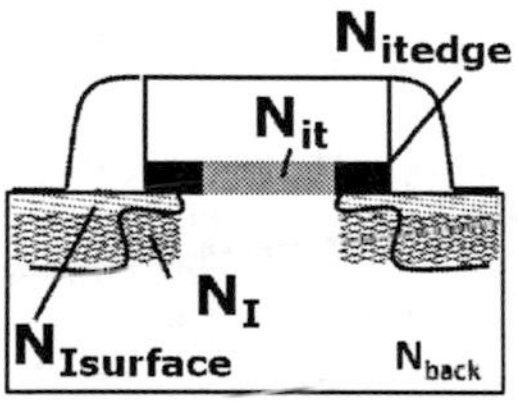
N_I N_{HighC}/N_{LowC}	1.35	
$N_{Isurface}$ (CET=5.2nm) N_{HighC}/N_{LowC}	10 to 1.3	
$N_{Isurface}$ (CET=2.3nm) N_{HighC}/N_{LowC}	/	
N_{it} N_{HighC}/N_{LowC}	1	
N_{itedge} (CET=5.2nm) N_{HighC}/N_{LowC}	1.3 to 3	
N_{itedge} (CET=2.3nm) N_{HighC}/N_{LowC}	1.3	

Monitoring the source/drain extension current, a carbon dose increase of a factor of 1.15 results in a defect density (N_I) ratio of 1.35 (table 4.7). The gate induced drain leakage indicates an increase of 1.3 in interface defects (N_{itedge}) for the thin gate dielectric (table 4.7). An increase in interface defect concentration (N_{itedge}) is also found for the PFETs with thick gate dielectric. Depending on the change of bulk defect concentration ($N_{Isurface}$), the ratio in interface defect density, due to the change in carbon dose, is between 1.3 and 3 (table 4.7) for this device.

Carbon co-implants are used to produce shallow junctions, and to avoid advanced annealing methods such as laser, or flash anneal. The main drawback of this technique is a carbon induced increase in defect density at the gate edge. Note, a SiO_x interlayer is present in todays high-k gate stacks. The rise in the defect density at the gate edge will become increasingly important with continued scaling. Defect density at the gate edge can be monitored, analyzing source/drain extension current, and gate induced drain leakage.

4.6 Arsenic and Phosphorus Implantation for Threshold Voltage Control

This section concentrates on the changes in electrical characteristics due to different halo- and V_{th} implants. Higher implantation doses are used to reach a lower subthreshold current, and a better short channel control. But higher implants also lead to an increase in SDE-, S/D-, channel-, and GIDL currents.

4.6.1 Description of Experiment

The arsenic V_{th} implant effects the doping concentration in the whole channel. The phosphorus halo implant is placed directly at the gate edge for short channel control [84]. The halo effects the threshold voltage at the gate edge. Changes of both implants lead to different electric field at the metallurgical junctions. These result in different leakage currents.

PFETs with three different phosphorus implantation doses in the halo: $2.1 \cdot 10^{13}$atm/cm^2 (Low Halo Dose), $2.3 \cdot 10^{13}$atm/cm^2 (Middle Halo Dose), and $2.5 \cdot 10^{13}$atm/cm^2 (High Halo Dose) are processed. Two different doses of the V_{th} implant in the channel are being investigated. An arsenic dose of $1.5 \cdot 10^{12}$atm/cm^2 (Low V_{th} Dose Thin), and $1.8 \cdot 10^{12}$atm/cm^2 (High V_{th} Dose Thin) are evaluated for the samples with thin gate dielectric. For the thick gate dielectric PFETs, an arsenic implant of $1.1 \cdot 10^{12}$atm/cm^2 (High V_{th} Dose Thick), and $1.0 \cdot 10^{12}$atm/cm^2 (Low V_{th} Dose Thick) are tested.

The samples are described more precisely in table B.2 and B.3 of the appendix. Figure 4.38 shows the SIMS profile of the MOSFET at the end of the process. The metallurgical S/D junction depth is between 90-100nm. No arsenic is detected at this depth. At the gate dielectric interface, both halo implant, and V_{th} implant are of equal importance.

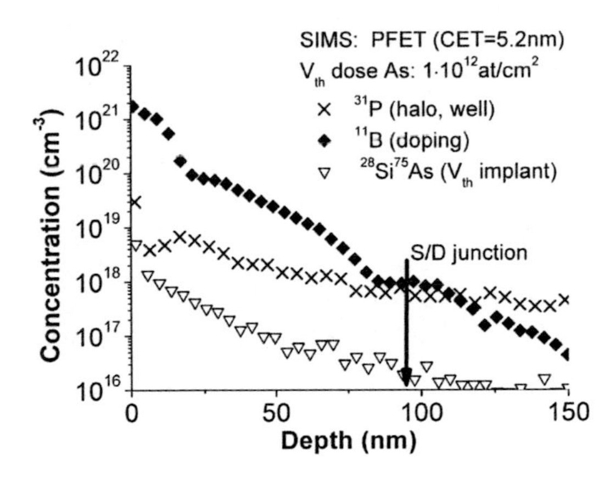

Figure 4.38: SIMS measurement of the PFET with low V_{th} implant, and a CET of 5.2nm at the end of processing [85]. The arsenic concentration is at detection limit at the metallurgical S/D junction. At the SDE junction both arsenic, and phosphorus implant are of equal importance.

4.6.2 Electrical Characterization of Transistor Performance

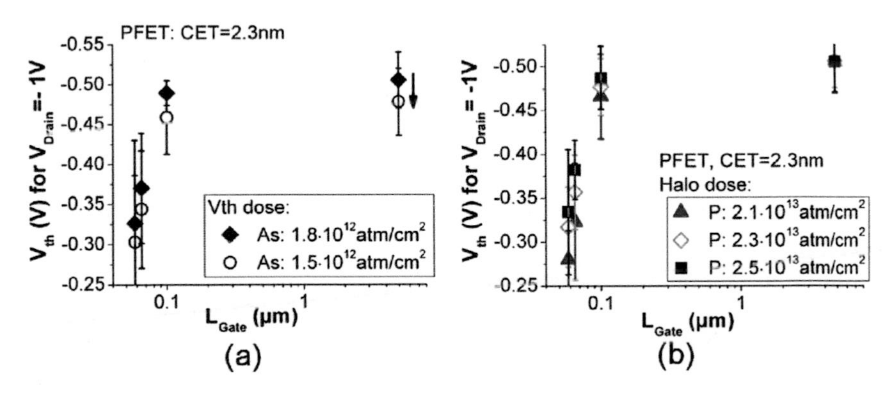

Figure 4.39: Roll off characteristics of the PFETs with a CET of 2.3nm at 85°C, for the devices with different:

(a) V_{th} implants.

(b) Halo implants.

Table 4.8: PFET Characteristics Depending on V_{th} Implant (L_{Gate}=65nm, CET=2.3nm)

	As: $1.5 \cdot 10^{12}$atm/cm^2	As: $1.8 \cdot 10^{12}$atm/cm^2
CET (nm)	2.33±0.01	2.32±0.02
L_{ov} (nm)	6.8±0.7	6.3±0.2
C_{Diode} (fF/μm^2) V_{Diode}=-1V (CV*)	1.27±0.01	1.28±0.02
V_{th} (V) V_{Drain}=-0.05V (CV**)	-0.35±0.05	-0.4±0.05
V_{th} (V) V_{Drain}=-0.05V (IV)	-0.46±0.01	-0.49±0.02
V_{th} (V) V_{Drain}=-1V (IV)	-0.34±0.07	-0.37±0.07
V_{fb} (V) (CV**)	0.9±0.05	0.9±0.05
I_{on} (μA/μm)	203.8±40.4	185.4±47.8
S_{Vth} (mV/dec) V_{Drain}=-0.05V	123.0±9.2	117.2±8.1
λ (IV***)	0.125±0.064	0.124±0.056

All values are given for measurements on 5 dies (wafer center).
* Capacitance voltage measurements are done on a diode structure with SDE and S/D implants.
** V_{fb} and V_{th} are taken from the point of maximum, and minimum slope of the CV characteristic.
*** DIBL parameter is determined after the V_{th} shift of $I_{Drain}V_{Gate}$ as described in section 2.2.

The effect of the change in V_{th} dose (table 4.8), and halo dose (table 4.9) on the basic electrical parameters is investigated in this subsection. A full comparison of the long channel thin-, and thick gate dielectric devices can be found in tables B.8 and B.9 of the appendix.

The PFETs with decreased arsenic dose show on average a lower V_{th}. The complete roll off curve is shifted downwards by a constant V_{th} of 0.03V for the transistor with thin gate dielectric (Fig. 4.39(a)). The overlap length is slightly increased with decreasing arsenic implantation dose (table 4.8). But the difference of the basic transistor parameters is small within the measurement statistics. Variations in basic transistor characteristics are not observed for the devices with thick gate dielectric. The capacitance of the S/D junction is not effected by the shallow channel implants.

The PFETs with an increase in halo dose show as expected a better short channel control, and improved roll off behavior (Fig. 4.39(b)). Also the overlap length slightly drifts to higher values with decreased halo dose (table 4.9). The changes are small below 1nm, and can not be observed in the long channel devices because of statistical deviation of the parameters. An increase in S/D capacitance with increasing halo dose is observed (table 4.9). The higher halo doping leads to a more abrupt junction with reduced space charge region width.

Table 4.9: PFET Characteristics Depending on Halo Implant (L_{Gate}=65nm, CET=2.3nm)

	P: $2.1\cdot10^{13}$atm/cm^2	P: $2.3\cdot10^{13}$atm/cm^2	P: $2.5\cdot10^{13}$atm/cm^2
CET (nm)	2.32±0.01	2.32±0.01	2.32±0.01
L_{ov} (nm)	7.0±0.1	6.5±0.8	6.2±0.6
C_{Diode} (fF/μm^2) V_{Diode}=-1V (CV*)	1.24±0.01	1.27±0.02	1.30±0.02
V_{th} (V) V_{Drain}=-0.05V (CV**)	-0.35±0.05	-0.37±0.08	-0.4±0.05
V_{th} (V) V_{Drain}=-0.05V (IV)	-0.45±0.02	-0.48±0.02	-0.49±0.01
V_{th} (V) V_{Drain}=-1V (IV)	-0.32±0.07	-0.36±0.04	-0.38±0.03
V_{fb} (V) (CV**)	0.9±0.05	0.9±0.05	0.88±0.08
I_{on} (μA/μm)	218.4±39.1	192.7±25.7	178.3±23.4
S_{Vth} (mV/dec) V_{Drain}=-0.05V	128.9±19.4	118.7±11.2	116.0±7.0
λ (IV***)	0.136±0.063	0.125±0.032	0.110±0.029

All values are given for measurements on 5 dies (wafer center).
* Capacitance voltage measurements are done on a diode structure with SDE and S/D implants.
** V_{fb} and V_{th} are taken from the point of maximum, and minimum slope of the CV characteristic.
*** DIBL parameter is determined after the V_{th} shift of $I_{Drain}V_{Gate}$ as described in section 2.2.

4.6.3 Leakage Current Dependent on Halo- and V_{th} Implant

Increased V_{th}-, and halo doses lead to a higher threshold voltage, and reduced subthreshold current. Also the source/drain-, source/drain extension-, and gate induced drain leakage are possibly effected. The changes of these leakage currents depending on the V_{th}-, and halo doses are investigated in this section.

Source/Drain Leakage

The S/D leakage depends on the defect density, and the electric field in the depletion region of the diode structure (Fig. 4.11(b)). The electric field is calculated from the CV characteristics (section 3.2). The V_{th} implant is below SIMS detection limit at a depth of about 90nm (Fig. 4.38). As expected, no effect of the V_{th} implant on the electric field, or leakage current in the S/D junction is found (Fig. 4.40(a)).

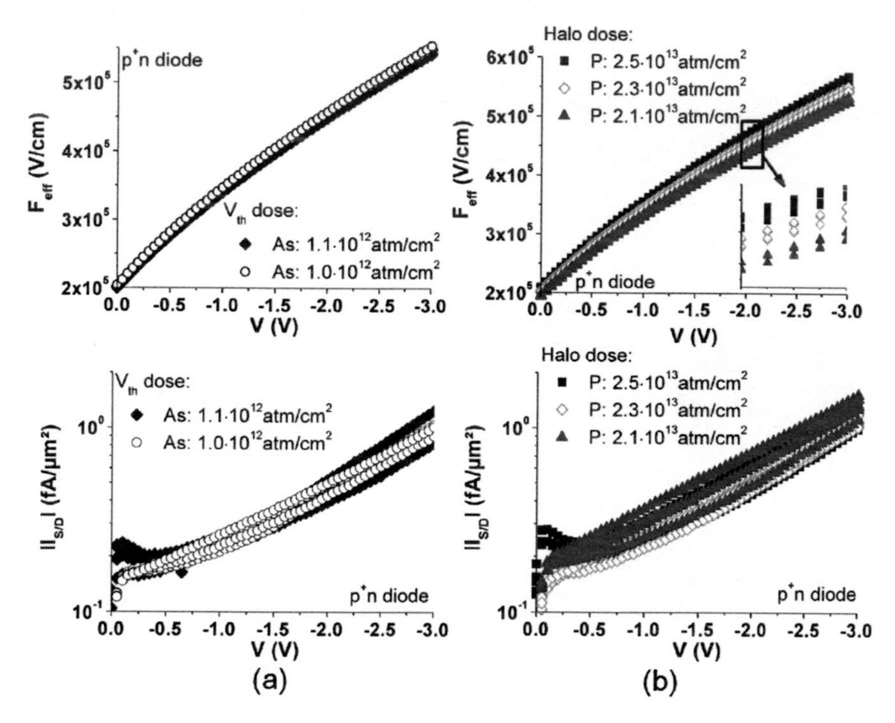

Figure 4.40: The upper part of the figure presents the electric field calculated from p^+n diode CV characteristics using the parameter r=0.4, and V_{int}=0.7 [43]. The resulting diode leakage current at 85°C is shown in the lower part. Five dies are measured for each sample. No difference in leakage current between the samples within the measurement variation is found. Diode structures of the thick gate dielectric PFETs are compared analyzing the effect of different:
(a) V_{th} doses.
(b) Halo doses.

Both halo, and well implant add up to n-doping in the PFET S/D depletion region. A small dependence of the electric field on the halo implant is measured (Fig. 4.40(b) upper part). The measured S/D current increase with the voltage is similar for all curves but the absolute value of the current varies (Fig. 4.40(b) lower part). The statistical variation by a factor of 1.6 of the absolute value of the leakage is due to changes in defect density. These variation in defect density is much higher than the influence of the electric field, so no clear dependence of the leakage current for a change in halo dose by a factor of 1.2 is observed.

Source/Drain Extension Leakage

The SDE leakage shows a strong increase with halo dose for the PFETs (Fig. 4.41(b) upper part, and Fig. B.10(b)). The simulations predict that the SDE leakage current is due to a trap band between 20-80nm (Fig. 4.13). The position of the metallurgical SDE junction is mainly caused by the phosphorus halo implant (Fig. 4.38). If the halo dose is raised by a factor of 1.2, the SDE leakage is increased by a factor of 1.9 at -3V for the PFET with thick gate dielectric, and a factor of 3.6 at -1V for the PFET with thin gate dielectric.

The arsenic concentration at a depth of 20nm below the silicon surface is by a factor of 10 less than the phosphorus concentration. So no dependence of the SDE leakage on a V_{th} dose variation of 1.2 is found in the measurement (Fig. 4.41(a) upper part, and Fig. B.10(a)).

Generation Leakage from Channel Region to Bulk

For the transistors with thin gate dielectric no dependence of the channel leakage on the V_{th} implant is observed (Fig. 4.41(a) lower part). The channel leakage of the PFET with thin gate dielectric is a generation current from the silicon to gate dielectric interface traps which is only slightly influenced by the electric field (section 4.4).

A strong increase in channel current by a factor of 1.6 for the thick gate dielectric device is measured (Fig. 4.41(a) upper part). No difference in average interface trap concentration between the devices with different V_{th} implants is found (Fig. B.11 of the appendix). In section 4.4, it is indicated that bulk traps close to the interface possibly contribute to the channel leakage. The bulk trap assisted channel leakage is not accurately simulated. The channel leakage is unchanged by the halo implantation dose (Fig. 4.41(b)).

Gate Induced Drain Leakage

For the devices with thin gate dielectric, no change in GIDL current for variation the implant dose variation within the measurement statistic is found (Fig. 4.42 lower part). The interface trap assisted GIDL current in these samples is caused by the band bending

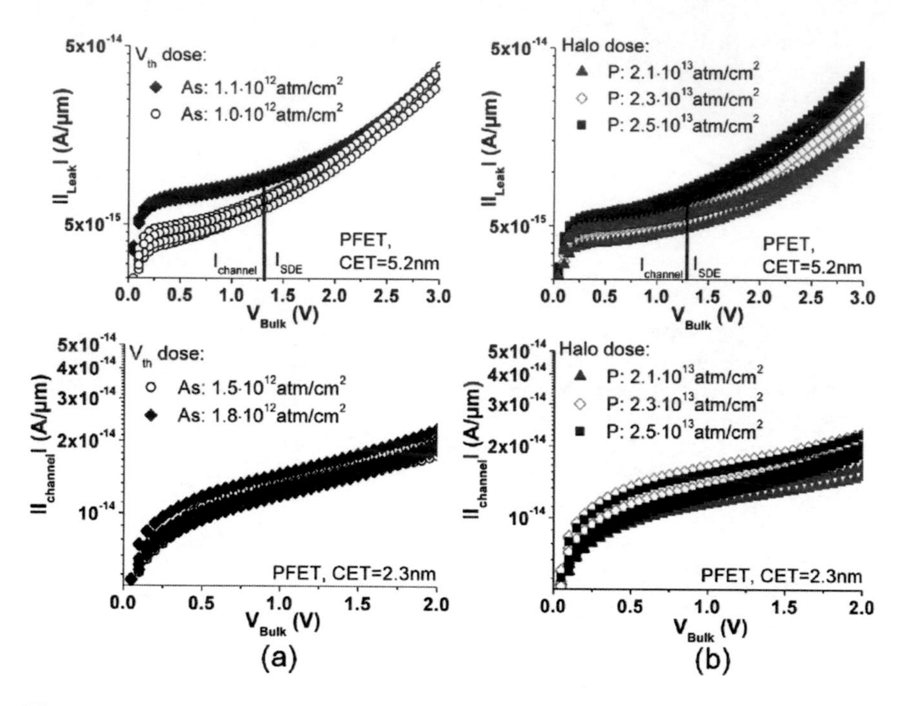

Figure 4.41: The upper part of the figure presents the channel-, and SDE leakage current of devices with a CET of 5.2nm. The channel leakage of transistors with a CET of 2.3nm is shown in the lower part. All measurements are done at PFETs with L_{Gate}=5μm at 85°C on five dies each. Different implant dose variations are compared:
(a) V_{th} implant.
(b) Halo implant.

between the gate dielectric, and the p^+-SDE region. So no effect by a factor 1.2 change in halo, and V_{th} implants is expected by the simulation (Fig. 4.43).

For the thick gate dielectric PFETs, the GIDL is caused by both bulk silicon traps close to the surface ($N_{Isurface}$), and interface traps (N_{itedge}) (Fig. 4.20(b)). The results are similar to the conclusions drawn from the SDE current analysis. An increase in halo implant leads to a rising GIDL, caused by a higher electric field which leads to an increasing current from the bulk traps, and from BTBT (Fig. 4.42(b) upper part). The GIDL at -3V is increased by a factor of 1.35 if the halo dose is increased by a factor of 1.2. The changes in V_{th} implant are too small to significantly effect the GIDL of the PFETs with thick gate dielectric.

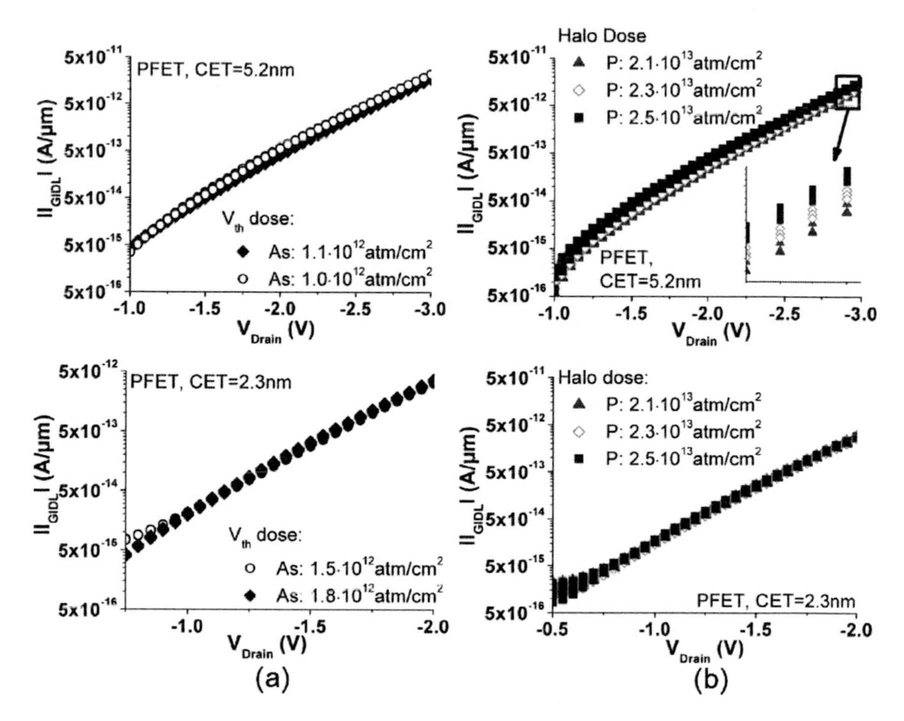

Figure 4.42: GIDL current for PFETs at 85°C on five dies each are presented. Measurements at devices with a CET of 5.2nm, and 2.3nm are shown in the upper, and lower part, respectively. Different implant dose variations are compared:
(a) Variations in the V_{th} dose do not effect the GIDL.
(b) Variations in halo dose lead to a higher GIDL for PFETs with a CET of 5.2nm.

Subthreshold Leakage

For 5μm long devices no exponential increase of the subthreshold current with the drain voltage is observed (Fig. 4.44 lower part, and Fig B.12). Hence, no DIBL occurs. As expected, an increased subthreshold leakage current is observed for the low V_{th} dose devices (Fig. 4.44). The halo implant does not effect the subthreshold current. For 5μm long PFETs the halo implant does not significantly change the important leakage parameters threshold voltage, and body factor.

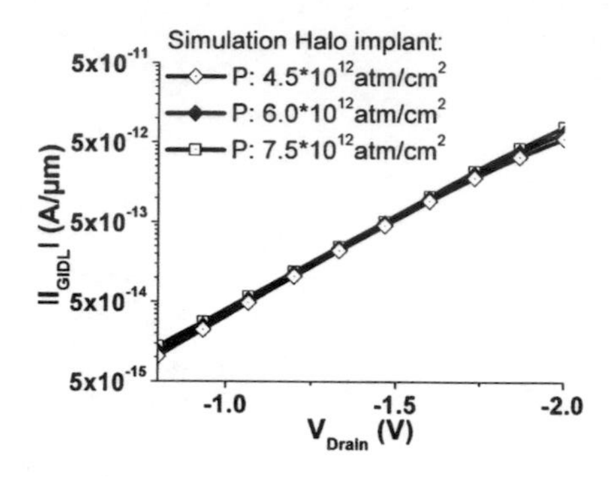

Figure 4.43: Simulation of the trap assisted GIDL current depending on the halo implant for PFETs with a CET of 2.3nm at 85°C.

Table 4.10: Parameters of Subthreshold Current from Extraction for PFETs (CET=2.3nm, L_{Gate}=65nm)

PFET	V_{th} (V)	λ	$^*\lambda_{measured}$	m	$m_{measured}$	$(\eta\epsilon_{Si})/(L_{Gate}\epsilon_{ox})$
As: $1.5\cdot10^{12}$atm/cm^2	0.46	0.125	0.125	1.9	1.73	0.0286
As: $1.8\cdot10^{12}$atm/cm^2	0.49	0.12	0.124	1.8	1.65	0.0290
P: $2.1\cdot10^{13}$atm/cm^2	0.45	0.13	0.136	1.95	1.82	0.0290
P: $2.3\cdot10^{13}$atm/cm^2	0.48	0.12	0.125	1.8	1.67	0.0290
P: $2.5\cdot10^{13}$atm/cm^2	0.49	0.115	0.11	1.75	1.66	0.0286

* Extracted by the V_{th} changes with drain voltages from the IV curves

The subthreshold current is enhanced by the DIBL effect at short gate lengths. The DIBL is dependent on halo, and V_{th} implants leading to an increasing subthreshold leakage with decreased implantation dose for the 65nm long PFETs (Fig. 4.44 upper part). The leakage controls the off-current of the thin gate dielectric transistor at -1V. Equation (4.1) is used to fit the leakage current with the body factor (m), and the DIBL constant (λ) as free parameters. The results are given in table 4.10. The geometry parameter (η), that is extracted analyzing λ, is expected to be constant (equation 2.5). No change in η for the different PFETs is found, so the fit is done correctly. Clearly, one can observe a decrease in DIBL constant due to the increased doping. Measured ,and fitted values are in good agreement.

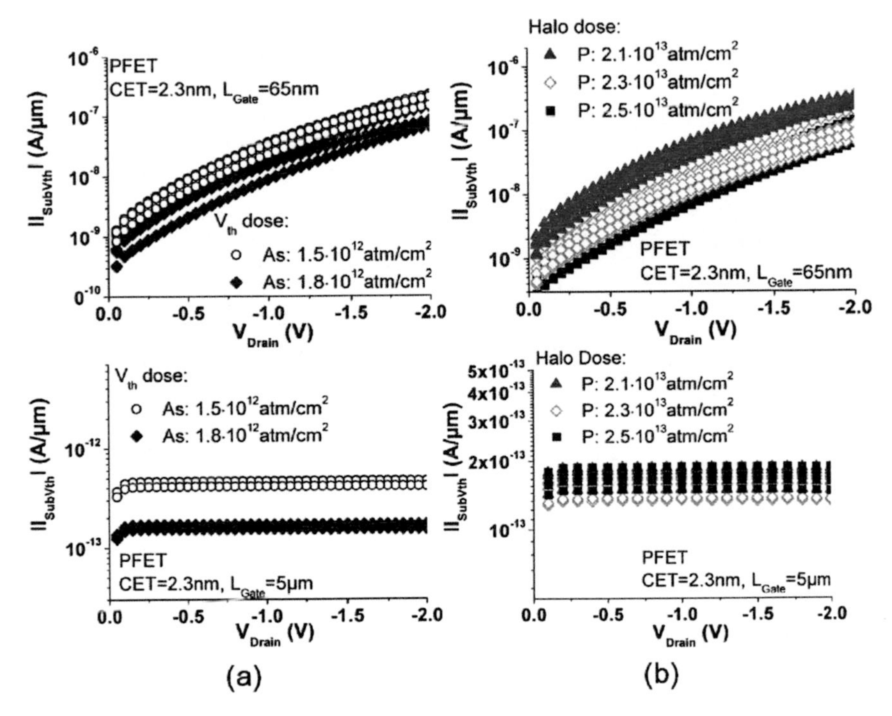

Figure 4.44: Subthreshold current at 85°C for five PFETs each with a CET of 2.3nm, and a channel length of 65nm (upper part), and 5μm (lower part). Different implant dose variations are compared:
(a) V_{th} implant.
(b) Halo implant.

4.6.4 Conclusion

In the last part of the chapter, the effect of different V_{th}-, and halo implants on the leakage current are under investigation. Higher implantation doses are used to reach a lower subthreshold current, and a better short channel control. But higher implants possibly lead to an increased SDE-, S/D-, channel-, and GIDL current. Table 4.11 compares the positive impact of increased doping concentration on the subthreshold current, with the negative effect on the other main leakage currents.

For the thin gate dielectric PFETs, a higher halo doping leads to a more abrupt junction, and to a higher source/drain extension leakage current. An increased halo doping also

Table 4.11: Measured Leakage Current Ratios for Halo- and V_{th} Dose Increase

CET=2.3nm		*Leakage Current Ratio at V_{Drain}=-1V and V_{Gate}=0V				
Implant	Dose (atm/cm^2)	GIDL	SDE	Channel	Gate	Subthreshold
V_{th}	$1.8\cdot10^{12}/1.5\cdot10^{12}=1.2$	1	1.6	1	1	1/2 (L_{Gate}=65nm) 1/2.7 (L_{Gate}=5μm)
Halo	$2.3\cdot10^{13}/2.1\cdot10^{13}=1.1$	1	1.6	1.1	1	1/2 1/1.1 (L_{Gate}=5μm)
Halo	$2.5\cdot10^{13}/2.1\cdot10^{13}=1.2$	1	3.6	1	1	1/5 1/1 (L_{Gate}=5μm)

CET=5.2nm		*Leakage Current Ratio at V_{Drain}=-3V and V_{Gate}=0V				
Implant	Dose (atm/cm^2)	GIDL	SDE	Channel**	Gate	Subthreshold
V_{th}	$1.1\cdot10^{12}/1.0\cdot10^{12}=1.1$	0.9	1	1.6	-	1/2.7 (L_{Gate}=5μm)
Halo	$2.3\cdot10^{13}/2.1\cdot10^{13}=1.1$	1.2	1.4	1.0	-	1/1.1 (L_{Gate}=5μm)
Halo	$2.5\cdot10^{13}/2.1\cdot10^{13}=1.2$	1.35	1.9	1.2	-	1/1.1 (L_{Gate}=5μm)

* Average measured leakage current ratio is given.

** Channel current ratio measured at -0.75V for thick dielectric PFETs.

effects the electric field in the source/drain region. But the change is small for the chosen variation in halo dose. No shift in source/drain leakage current is observed, that is not within the variation of the defect density. The channel leakage from the gate dielectric to silicon interface traps is not effected by neither the V_{th}-, nor the halo dose. The gate induced drain leakage current in the thin gate dielectric device is due to interface traps at the silicon to spacer oxide, and silicon to gate edge dielectric. The gate induced drain leakage is not effected. The current arises in the overlap region, and is dependent only on boron doping in the SDE [50]. The gate leakage current is not changed for different halo-, or V_{th} doses. The main leakage current at a bias of -1V, equivalent to operating conditions, for the short channel devices, is subthreshold current and gate leakage. The highest implantation doses give the best results in terms of leakage for these devices.

For the devices with thick gate dielectric, a small increase in V_{th} dose by a factor of 1.1 leads to a high rise in channel leakage. It is possible that the contribution of silicon bulk traps to the channel leakage is underestimated. The gate induced drain leakage, and the source/drain extension current are similarly increased with halo dose. The trap assisted leakage, and BTBT current at the SDE junction rises, due to the increased n-doping of the PFET SDE depletion region. The main leakage current at a bias of -3V, equivalent to operating conditions, for long and short channel devices is the gate induced drain leakage. The lowest halo dose gives the best results in terms of leakage for these devices.

Leakage currents need to be reduced further for low standby power targets. To get the best results, the main leakage current under operating condition is determined as shown in this study, then the optimization of source/drain implants is possible.

5 Impact of High-k Process Adjustment on Transistor Leakage Current

In order to have a good control over the transistor channel, high permittivity dielectrics are used resulting in a low capacitance equivalent thickness (CET) of the gate stack, and a low gate leakage current. Different adjustments have to be made to integrate the high-k dielectric in the transistor process flow. In the following chapter, the effects of important process adjustments on the leakage currents of high-k transistors are investigated.

5.1 High-k Process Flow

The peripheral DRAM transistors are prepared using Qimonda's 65nm process flow. The gate stack is presented in figure 5.1. The NFET has a high-k - polysilicon gate. A high-k - metal gate stack is used for the PFET. Due to the metal electrode, no polysilicon depletion occurs, and the CET is decreased. The most relevant process steps are described in detail in this section (Fig. 5.2, and 5.3).

In the first step, a shallow trench isolation is formed using silicon oxide. Phosphorus, and boron are implanted in the PFET, and NFET well, respectively. A shallow implant for threshold voltage control follows. Arsenic is used for the p-channel transistors, and boron for the n-channel devices. This implant will be called V_{th} implant in the following.

The gate oxide is processed in the following steps. DRAM peripheral transistors with two different gate oxides are investigated. The so called "thick gate dielectric" transistors target a CET of 5.5nm, and the "thin gate dielectric" MOSFETs target a CET of 2nm. For the PFET with thick gate oxide, a 4nm silicon oxide interlayer is grown by a thermal oxidation prior to the high-k deposition. For the thin gate oxide devices, the thermal oxide is removed by wet etch. A 1nm silicon oxide interlayer grows during wet chemical processing.

In the next step, a hafnium silicon oxide of about 2.5nm is deposited by atomic layer deposition (ALD), followed by a rapid thermal anneal (RTA) at 1000°C. After the formation of the dielectric, the NFET gate electrode is processed. About 20nm undoped polysilicon is deposited on top of the dielectric. A 15nm silane oxide hard mask is grown to protect the NFET electrode material during PFET processing.

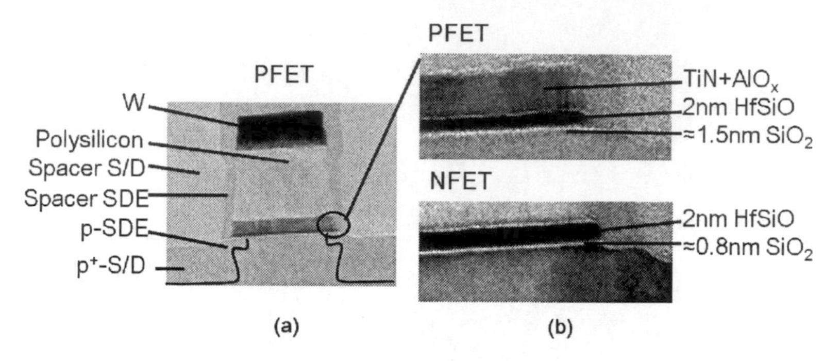

Figure 5.1: Transmission electron microscopy (TEM [112]) of the fabricated MOSFETs is presented.
(a) Overview of the PFET with a silicon nitride spacer is shown.
(b) Magnification of the gate stack is presented. The silicon oxide interface has a thickness of about 1nm.

The hard mask, and the polysilicon are etched from the PFET area (Fig. 5.2(7)), and a few atomic layers of aluminum oxide are deposited by ALD. The aluminum is diffused into the gate dielectric by a RTA at 1000°C. This process step is important to reach the threshold voltage target of the PFET [113]. The PFET metal electrode of 8nm titanium nitride is deposited by physical vapor deposition (PVD). 10nm undoped polysilicon is grown on top of the metal gate electrode to ensure about the same final height of NFET, and PFET gate stack

The metal, and aluminum oxide are removed from the NFET gate stack by etching. A chemical mechanical polishing procedure is used for planarization of the NFET, and PFET polysilicon top layers. The polysilicon surface is cleaned, and subsequently a layer of about 40nm amorphous silicon is deposited. The silicon electrode is doped using a phosphorus implantation for both NFET, and PFET. The phosphorus is electrically activated, and diffuses towards the lower polysilicon layer by an RTA step at 1000°C. The n^+ polysilicon acts as gate electrode for the NFET. The gate stack is completed by sputtering a stack of 3nm titanium, 7nm tungsten nitride, 13nm tungsten, and a silicon nitride capping layer. An RTA forming gas anneal at 850°C is done after the tungsten deposition.

For gate patterning, the gate electrode is etched from the source/drain area using a combination of dry, and wet etch processes (Fig. 5.3(15)). In order to remove the hafnium silicon oxide, a germanium ion implantation with an energy of 10keV, and a dose of $3.0 \cdot 10^{14}$ atm/cm^2 is used to preamorphize the dielectric layer. This step is necessary to completely remove the hafnium silicon oxide [114]. The amorphous high-k, and silicon oxide are removed from the source/drain area by a wet etch (Fig. 5.3(17)).

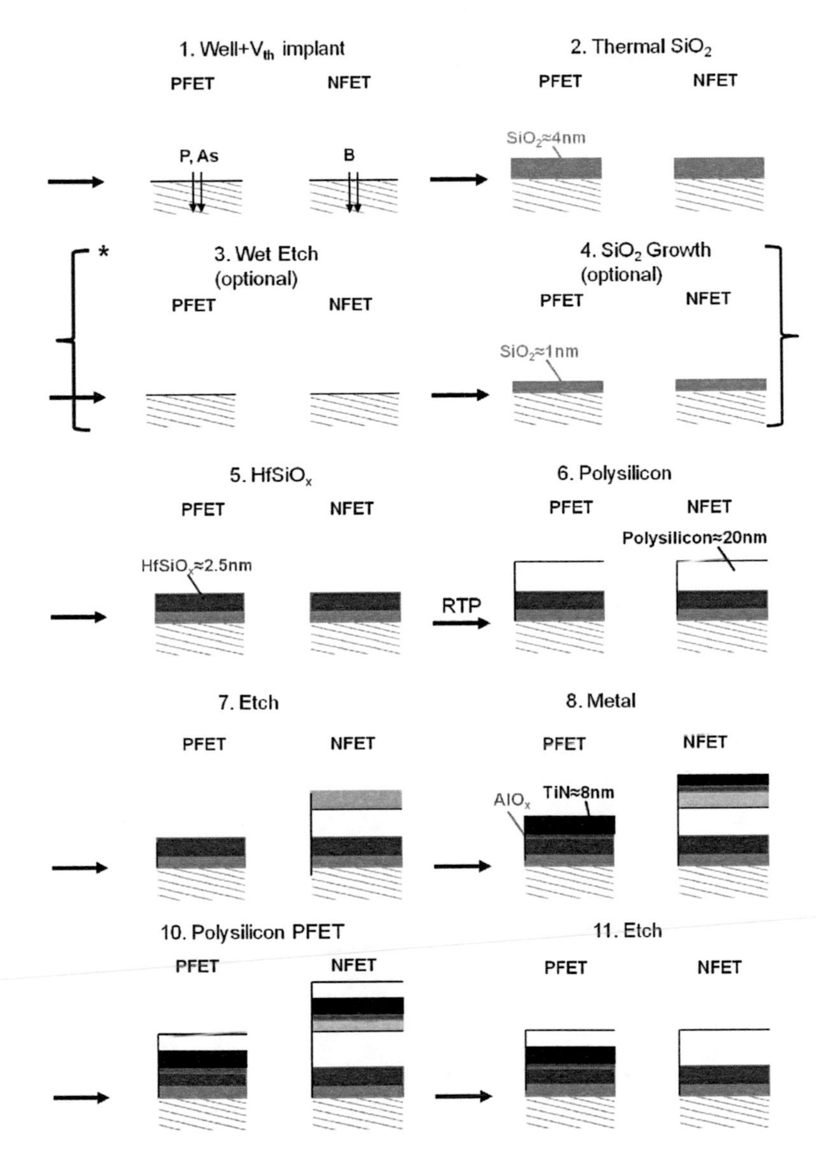

Figure 5.2: Most important process steps of the MOSFET fabrication for the high-k devices are presented (shaded: Si substrate, gray: SiO_x, white: Polysilicon, amorphous Si, red: $HfSiO_x$, violet: AlO_x, black: metal).
* Dual gate oxide process: Indicated process steps are used only for PFETs with thin gate dielectric.

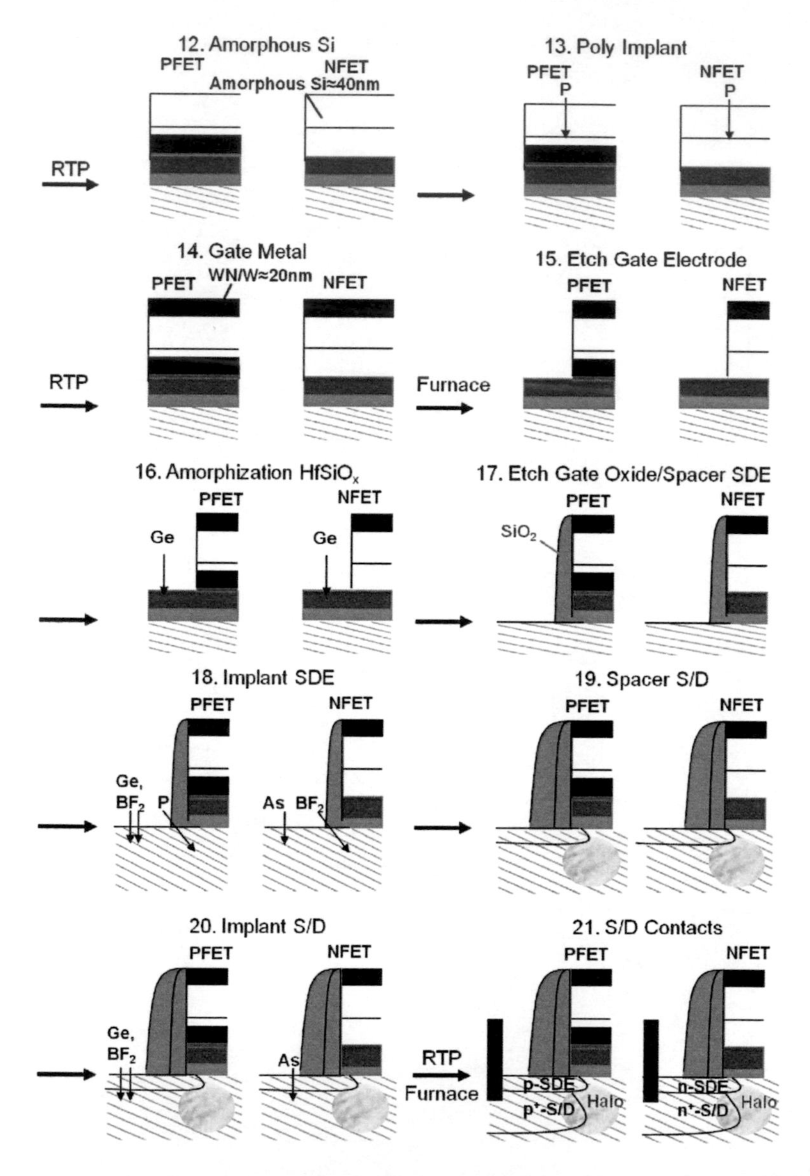

Figure 5.3: Most important process steps of the MOSFET fabrication for the high-k devices are presented (shaded: Si substrate, gray: SiO_x, white: Polysilicon, amorphous Si, red: $HfSiO_x$, violet: AlO_x, black: metal).

A silicon oxide, or silicon nitride spacer of about 8nm is deposited, and etched in the next process step. The spacer protects the gate edge during the following ion implantation steps, and defines the overlap length between the source/drain extension (SDE) and the gate.

A shallow SDE implant ensures a small overlap between the source/drain region and the gate [77]. The NFET extension is formed by an arsenic implant for the thin gate dielectric devices, and a phosphorus plus arsenic implant for the transistors with thick gate dielectric. The silicon substrate of the NFETs is amorphized by the doping implantation, because of the high arsenic atomic mass. The NFET halo is implanted in the next process step.

Halos are implanted to increase channel doping concentration at the gate edge. Those implants are known to reduce short channel effects [84]. A tilt of 28° is used for the quadruple implantation. The implantation also effects the threshold voltage in the short channel devices. Boron diflouride is implanted in the NFET halo, followed by a 1000°C RTA anneal to activate of the dopants, and recrystallize the silicon.

Two implantation steps are necessary to form the source/drain extension of the PFETs. A germanium implant is used for an amorphization of the silicon substrate. The germanium implantation prevents the channeling effect during the subsequent boron diflouride doping implantation [78]. The germanium amorphization introduces a high amount of interstitials [79]. Phosphorus is implanted to form the PFET halo. The PFET substrate is recrystallized during the subsequent spacer formation.

Prior to the formation of the source/drain, a silicon oxide spacer of about 27nm is deposited from a tetra-ethyl-ortho-silicate precursor (TEOS), and is etched. An arsenic implantation step completes the source/drain junction of the NFET devices. For the PFETs, a germanium amorphization, and boron diflouride is implanted to form the source/drain.

Table 5.1: Implantation Energies, and Doses for the High-k Devices (CET≈2nm)

Implant	Species	PFET		Species	NFET	
		[keV]	[atm/cm^2]		[keV]	[atm/cm^2]
Well	P	340	$5.3 \cdot 10^{13}$	B	150	$5.3 \cdot 10^{13}$
V_{th}	As	30	$9 \cdot 10^{11}$	B	10	$7 \cdot 10^{12}$
Poly	P	7	$5 \cdot 10^{15}$	P	7	$5 \cdot 10^{15}$
SDE	Ge	15	$3 \cdot 10^{14}$			
	BF_2	3	$3 \cdot 10^{14}$	As	3	$1 \cdot 10^{15}$
Halo	P	40	$1.8 \cdot 10^{13}$	BF_2	45	$1.4 \cdot 10^{13}$
S/D	Ge	15	$3 \cdot 10^{14}$			
	BF_2	10	$2.5 \cdot 10^{15}$	As	16	$3 \cdot 10^{15}$

The dopants are activated using two RTA process steps at 1000°C. A final thermal process step is required in DRAM processing to reduce the defect density, and improve cell

Table 5.2: Implantation Energies and Doses for the High-k Devices (CET≈5.5nm)

Implant	Species	PFET		Species	NFET	
		[keV]	[atm/cm^2]		[keV]	[atm/cm^2]
Well	P	340	$5.3{\cdot}10^{13}$	B	150	$7.0{\cdot}10^{12}$
V_{th}	As	30	$5{\cdot}10^{11}$	B	10	$1.8{\cdot}10^{12}$
Poly	P	7	$5{\cdot}10^{15}$	P	7	$5{\cdot}10^{15}$
SDE	Ge	15	$3{\cdot}10^{14}$			
	BF_2	3	$3{\cdot}10^{14}$	P	8	$2.5{\cdot}10^{13}$
				As	8	$1.2{\cdot}10^{14}$
Halo	P	40	$1.8{\cdot}10^{13}$	BF_2	60	$1.5{\cdot}10^{13}$
S/D	Ge	15	$3{\cdot}10^{14}$			
	BF_2	10	$2.5{\cdot}10^{15}$	As	16	$3{\cdot}10^{15}$

retention time. This DRAM anneal is a 30min furnace anneal at 800°C. The source/drain contacts are formed by a titanium/ titanium nitride liner, and tungsten filling.

The implantation energies and doses for PFET, and NFET are given in tables 5.1 and 5.2. The total halo dose of the quadruple implantation is given in the tables. Secondary ion mass spectroscopy (SIMS) doping profiles of the source/drain junctions are shown in figure 5.4. The geometry parameters of the measured transistors can be found in table A.2.

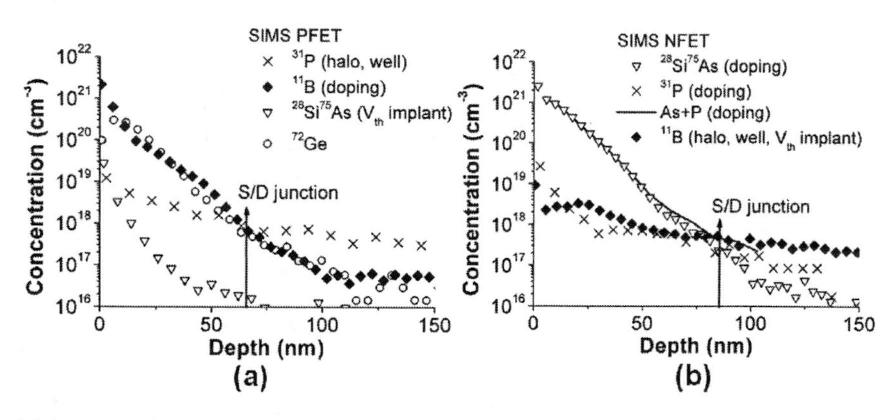

Figure 5.4: SIMS profile of the S/D junction after processing for MOSFETs with a CET of about 5.5nm [85] is presented for a:
(a) PFET
(b) NFET

This chapter focuses on the effect of the high-k process flow on the leakage currents of the transistors. In the first part of the chapter, the intrinsic leakage mechanisms for the high-k transistors are explained. The leakage currents are compared to a standard silicon oxynitride MOSFETs based on the 65nm technology (table C.4). Besides the changes in dielectric material, and gate stack processing, the silicon oxynitride transistors require higher halo-, and V_{th} doping concentration to reach the transistor performance targets. The gate oxide to silicon interface quality of the devices is also analyzed.

Table 5.3: Sample Description for Transistors of the 65nm Technology

Sample Set	Experiment	Additional Information
SDE spacer set	SiO_2 SDE spacer	10keV Ge preamorphization
	SiN SDE spacer	10keV Ge preamorphization
Etch set	10keV Ge	SiN SDE spacer improved gate electrode etch
	3keV Ge	SiN SDE spacer improved gate electrode etch
SiON	Reference	SiO_2 SDE spacer no Ge preamorphization

The effects of the SDE spacer material, and germanium preamorphization for gate patterning, on the leakage currents are investigated in the second part of the chapter. Two high-k sample sets are prepared (table 5.3). Silicon nitride, and silicon oxide as SDE spacer materials are compared in the first sample set, called "SDE spacer set". The germanium preamorphization implantation energy prior to the $HfSiO_x$ etch is varied between 3keV and 10keV in the second sample set, called "etch set". Both sample sets differ slightly in thermal budget, and etching chemistry (Fig. 5.5).

5.2 TCAD Simulation of the Process

TCAD simulations are performed to understand the effect of the changes in process flow on the defect density, and the electric field. A thin p-channel transistor is chosen for device simulation. The main production steps of the PFET device are simulated with TCAD Sentaurus process. The temperature budget is adjusted to reproduce the doping profile measured by SIMS analysis.

Figure 5.6 shows a comparison of the simulated, and measured SIMS doping profiles of the source/drain along the cut line. For the boron doping profile, a reasonably good agreement is achieved. The measured halo doping concentration does not match the simulation. This is caused by the fact that the halo implantation dose is changed in the simulation to match the drain induced barrier lowering (DIBL) characteristics (Fig. 5.7). SIMS measurements are done on a diode structure. In the diode structure the quadruple phosphorus implant is not blocked at the gate stack, therefore a slightly different halo doping profile is obtained.

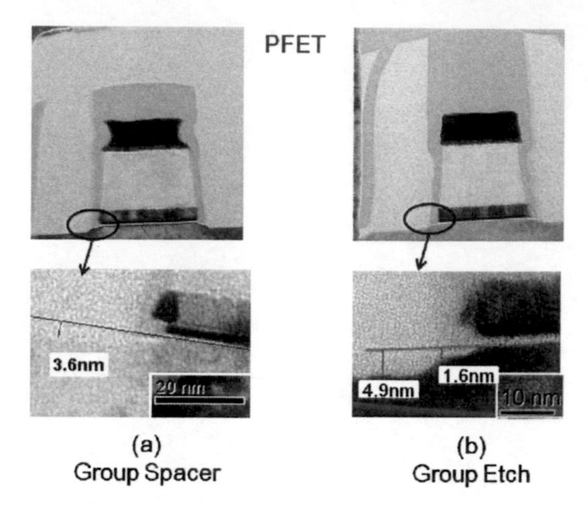

Figure 5.5: TEM analysis of the PFETs of the two sample sets showing the results of the variation in the etching chemistry:
(a) SDE spacer set [112].
(b) Etch set [115].

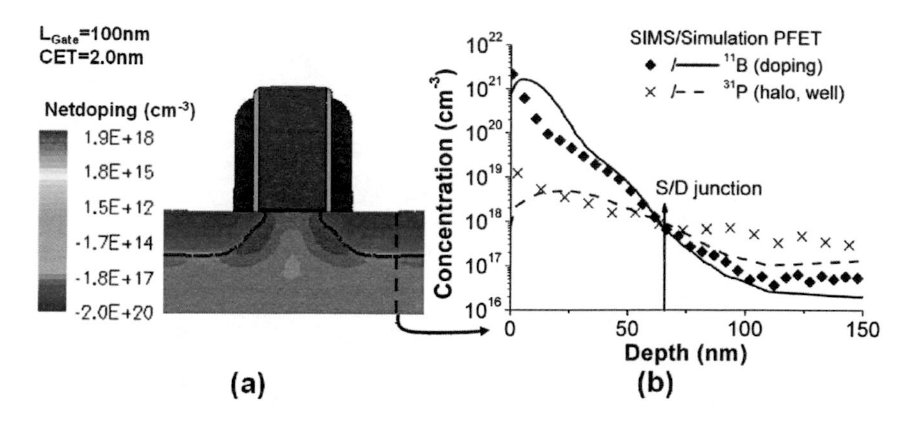

Figure 5.6: Simulated, and measured doping profiles are shown.
(a) Transistor structure of a 100nm long PFET (CET=2.0nm) from process simulations
(b) SIMS profile of the S/D junction after processing [85] is presented. The process simulation gives a reasonable agreement with the measured doping gradients.

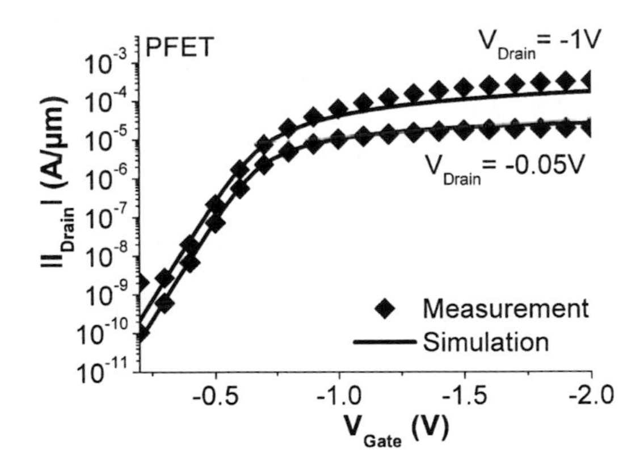

Figure 5.7: Subthreshold characteristic at different drain voltages for the PFET of the sample set with different SDE spacers (SiO_2 SDE spacer) is presented. A good agreement between measured, and simulated curves is reached for PFETs with a gate length of 100nm, and a CET of about 2nm at 85°C.

5.3 Interface Traps

The interface trap concentration at the gate dielectric to silicon interface is measured by the charge pumping (CP) technique. Table 5.4 lists the measured results of the average trap concentration per area (N_{it}). The interface defect density of the high-k PFET with thin gate dielectric is increased by a factor of three compared to the thick gate dielectric PFET. The results of the high-k NFET show the opposite trend. The defect concentration is decreased by a factor of two with decreasing interlayer thickness.

The silicon to gate dielectric interface of the high-k PFET with thick gate dielectric has approximately the same quality as the comparable silicon oxynitride technology (table 5.4). The high-k NFET with thick gate dielectric shows a clear degradation compared to the silicon oxynitride sample (table 5.4).

All high-k NFET devices have a reduced interface defect concentration compared to the high-k PFETs. One possible explanation is that the aluminum from the threshold voltage tuning layer degrades the silicon to gate dielectric interface [113]. Literature also suggests the formation of oxygen deficient traps due to the titanium nitride top electrode [116]. The titanium nitride top electrode, and aluminum oxide V_{th} tuning layer are placed closer to the interface for the thin gate oxide PFETs. The increased influence of the V_{th} tuning layer, and/or metal electrode explains the higher N_{it} of the thin- compared to the thick gate dielectric PFETs.

Table 5.4: Si/Gate Dielectric Interface Defect Concentration for Different Sample Sets

Sample Set		N_{it}^{*} (cm^{-2})			
	Device	PFET		NFET	
		Thin	Thick	Thin	Thick
Reference	SiON	$(1.3\pm0.2)\cdot10^{10}$	$(1.3\pm0.2)\cdot10^{10}$	$(4.7\pm1.6)\cdot10^{9}$	$(2.4\pm0.5)\cdot10^{9}$
Etch set	Ge 10keV	$(4.7\pm1.2)\cdot10^{10}$	$(1.6\pm0.2)\cdot10^{10}$	$(3.9\pm0.6)\cdot10^{9}$	$(10.0\pm0.7)\cdot10^{9}$
	Ge 3keV	$(5.0\pm0.7)\cdot10^{10}$	$(1.5\pm0.2)\cdot10^{10}$	$(4.2\pm0.3)\cdot10^{9}$	$(9.0\pm0.2)\cdot10^{9}$
SDE Spacer	SiN	$(7.3\pm2.1)\cdot10^{10}$	$(2.5\pm0.2)\cdot10^{10}$		
set	SiO_2	$(7.2\pm0.7)\cdot10^{10}$	$(2.3\pm0.1)\cdot10^{10}$		

*The interface trap density is determined by CP sweeps at the maximum of the base sweep at 1MHz, 1.5V amplitude, and 100ns rise- and fall time.

Energy dependent CP scans are preformed to get a better insight into the energetic levels of the defects. The measurements are done according to the method described in section 3.3. The change of defect concentration with the trap energy is presented in figure 5.8. The results agree well with published data [117]. For the PFET devices a defect band occurs at an energy range of 0.20eV to 0.25eV above mid gap. Traps at the conduction band edge in aluminum gated MOSFETs have been observed before [118]. Defect levels at the conduction band edge have been also assigned to oxygen vacancies [119]. The energy dependent sweeps of the high-k NFET samples show an increase of defect concentration at the conduction band edge, which strongly increases for the thick gate dielectric transistors (Fig. 5.8(b)).

Transistors with a gate length of 0.8μm are used for the CP analysis (see also table A.2). The N_{it}, measured at long channel transistors, is not effected by different SDE spacer materials, or variations of the germanium implantation energy during gate patterning (table 5.4). The CP current is directly proportional to the measured area. The CP current from the interface defects at the gate edge is not be detectable within the measurement variations of the total CP current.

The etch sample set shows an improved interface defect density compared to the sample set with different SDE spacers (table 5.4). The improvement is probably caused by the variations in the etch processes, and the difference in thermal budget between the two sample sets. The change in gate dielectric to silicon interface defect density between the two high-k sample sets can not be observed in the mobility characteristics (Fig. 5.9).

In conclusion, the charge pumping measurements reveal the degradation of the PFET silicon to gate dielectric interface due to the top electrode, or the threshold voltage tuning layer. This degradation is reduced if the SiO_2 interlayer gets thicker. The sample set with different gate etches has a reduced interface defect density compared to the set with different SDE spacers. The improvement, probably due to changes in thermal budget and etching process, does not lead to enhanced mobilities.

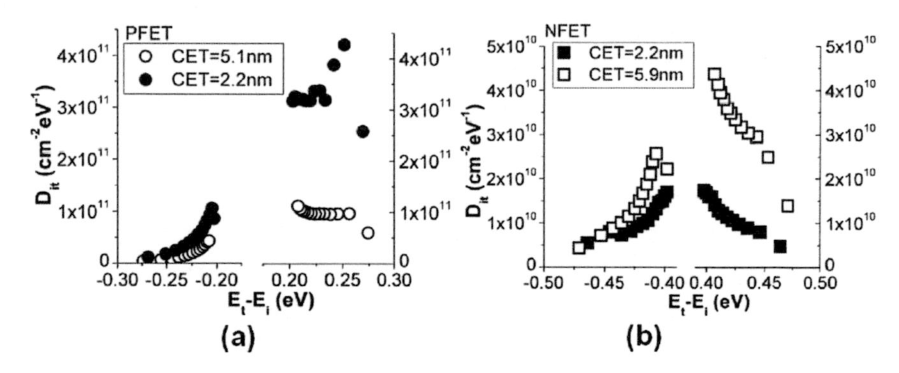

Figure 5.8: D_{it} determination by energy dependent CP measurements with an amplitude of ±1.5V at 25°C for typical transistors of the etch sample set is shown for a:
(a) PFET, with capture cross sections of $1.1{\cdot}10^{-14}$cm^2, and $9.3{\cdot}10^{-15}$cm^2 estimated for the thin, and thick gate dielectric transistor, respectively.
(b) NFET, with a capture cross section independent of the gate dielectric of $5.6{\cdot}10^{-18}$cm^2.

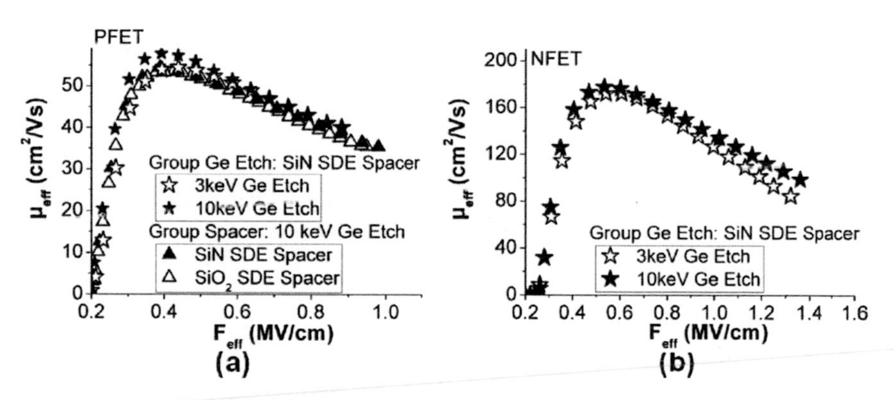

Figure 5.9: Mobility versus electric field measured at 85°C. The median measured curve of five transistors (L_{Gate}=10μm, CET≈2nm) is presented.
(a) PFET mobility depending on processing
(b) NFET (NFET 2 table B.4) mobility depending on processing

5.4 Electrical Measurement and Simulation of Leakage Current

Off-current leakage occurs when the gate is biased to 0V. Different voltage conditions at the source/drain, and the bulk lead to different leakage paths. Two main cases for peripheral DRAM transistors are presented: Either source, and drain are shorted when the voltage is ramped, or only drain potential is ramped while source is grounded (Fig. 5.10).

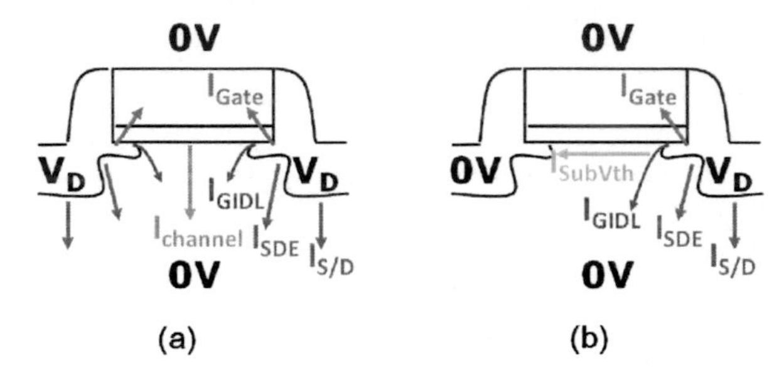

Figure 5.10: Schematic of the measured leakage currents occurring in off-state (V_{Gate}=0V), when:
(a) $V_{Source}=V_{Drain}$.
(b) V_{Source}=0V.

The results presented in Fig. 5.11, and Fig. 5.12 are for the high-k MOSFETs with the best performance and lowest leakage: The transistor with a 3keV germanium implant energy, and silicon nitride spacer from the sample set with different etches. Equivalent results are obtained for the MOSFETs from the other sample set (table 5.5). Detailed graphs can be found in section C.2 of the appendix. Figure 5.11 shows the different leakage current contributions for the 100nm long thin gate dielectric PFET, and NFET when the source is grounded. The dominant leakage current contributions, of the total transistors leakage, depending on voltage is shown in the bar graph of the figure 5.12. TCAD simulations are used to analyze the defect distribution, and leakage current mechanisms of the PFETs in detail. In this section, the fitted parameters of the median measurement curves are given (section C.4 of the appendix).

The conditions equivalent to the operation cases for peripheral DRAM MOSFETs are -1V at the drain for the thin gate dielectric transistor, and -3V at the drain for the devices with thick gate dielectric. The detailed leakage curves of the devices with thick gate dielectric can be found in section C.2 of the appendix. The main leakage current mechanism, at DRAM operation conditions, is the gate induced drain leakage (GIDL) for the high-k PFETs. The GIDL is also the main off-current of the high-k NFETs with thick

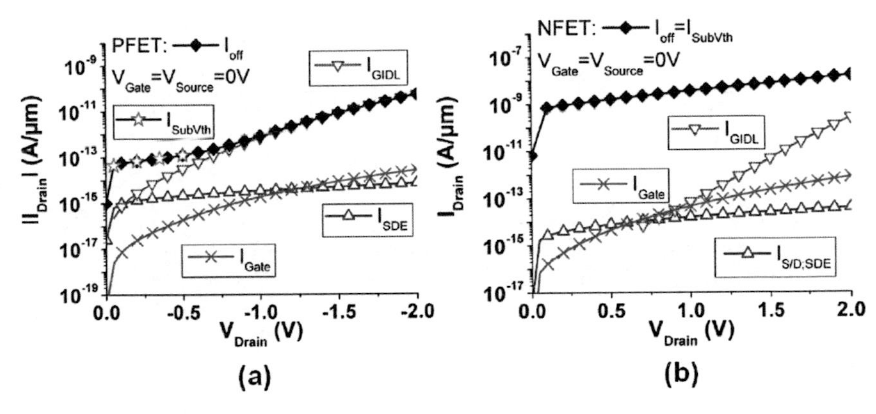

Figure 5.11: Typical leakage current voltage plot depending on the drain voltage for the 100nm long MOSFET with a CET of about 2nm at 85°C. The samples with the best performance from the etch sample set with a 3keV germanium implant are shown:
(a) PFET.
(b) NFET.

gate dielectric. Gate leakage, and GIDL add up to the overall off-current of the high-k NFETs with thin gate dielectric if source, and drain potentials are shorted. When the source is grounded, the subthreshold current of the 100nm long devices makes up 99.9% of the total off leakages for high-k NFETs with thin gate dielectric.

Table 5.5: Leakage Contributions to the Total Off-Current at ±1V for 100nm Long MOSFETs*

Device		$I_{S/D,SDE}$	I_{Gate}	I_{GIDL}	I_{SubVth}
		[pA/µm]			
NFET	SiON	$8.1\cdot10^{-2}$	$1.0\cdot10^{-1}$	$3.6\cdot10^{-1}$	$2.0\cdot10^{2}$
	3keV Ge (Etch set)	$9.9\cdot10^{-3}$	$1.2\cdot10^{-2}$	$1.9\cdot10^{-2}$	$2.5\cdot10^{3}$
	10keV Ge (Etch set)	$1.6\cdot10^{-2}$	$2.3\cdot10^{-2}$	$7.5\cdot10^{-2}$	$4.6\cdot10^{3}$
PFET	SiON	$6.7\cdot10^{-4}$	$4.0\cdot10^{-2}$	$3.5\cdot10^{-2}$	$1.4\cdot10^{4}$
	3keV Ge (Etch set)	$3.5\cdot10^{-3}$	$9.4\cdot10^{-4}$	$5.8\cdot10^{-1}$	$1.9\cdot10^{-1}$
	10keV Ge (Etch set)	$2.5\cdot10^{-2}$	$5.4\cdot10^{-4}$	$1.0\cdot10^{1}$	$1.4\cdot10^{-1}$
	SiN (SDE Spacer set)	$5.6\cdot10^{-2}$	$1.8\cdot10^{-3}$	$2.1\cdot10^{0}$	$3.5\cdot10^{-1}$
	SiO_2 (SDE Spacer set)	$7.3\cdot10^{-2}$	$1.7\cdot10^{-5}$	$7.0\cdot10^{-1}$	$5.5\cdot10^{1}$

*Measurement conditions: Source and gate are biased to 0V and a drain potential is applied. Average value of five dies is given.

For the thin silicon oxynitride reference devices, the gate leakage has as expected a high

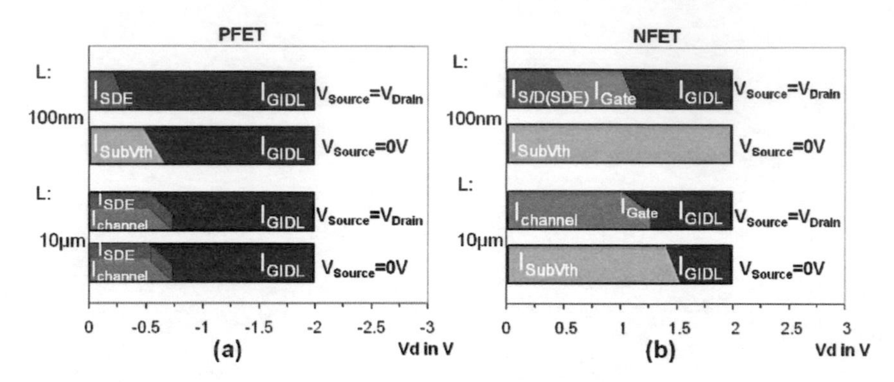

Figure 5.12: The bar graph shows the main leakage current contribution to the total off-current for short, and long channel MOSFETs with a CET of about 2nm depending on the drain voltage at 85°C. The samples with the best performance from the etch sample set with a 3keV germanium implant are shown:
(a) PFET.
(b) NFET.

influence on the total off-current. The subthreshold leakage is a main contribution to the off-current of the thin gate dielectric MOSFETs when the source is grounded. For all thick gate dielectric devices, the GIDL governs the off-current under operation conditions (see also section C.2 of the appendix). All leakage currents will be analyzed in more detail in the next sections.

5.4.1 Source/Drain Leakage

The source/drain leakage ($I_{S/D}$) occurs at the vertical diffused source/drain junction (Fig. 5.13(a)). A diode test structure is used to measure the S/D current. The advantage of the diode test structure is that the effective electric field can be determined from capacitance voltage measurements (section 3.2). Knowing the electric field makes it possible to analyze the dominant leakage current mechanism.

The voltage dependence of the S/D leakage current is given in the upper part of figure 5.14. The activation energy (E_a) is calculated from the temperature dependent S/D leakage using equation (2.2). The calculated activation energy decreases with the voltage (Fig. 5.14 lower part).

Figure 5.14 compares the measurements for p^+n- and n^+p diodes with the current calculated from the analytical approximation of the BTBT-, generation-, and Hurkx current (equations 2.7, 2.11, 2.12) [43]. The Hurkx current is the main leakage mechanism of

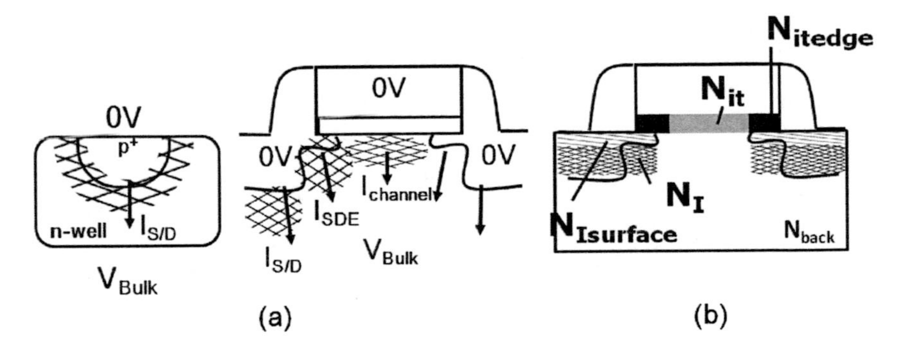

Figure 5.13: The Schematic sketch of the:
(a) Leakage current occurrence in the MOSFET is shown.
(b) Trap distribution of the transistors assumed in the simulation is presented.

the diodes. The fitting parameter of the effective mass determines the slope of the leakage with the voltage. In analytical calculation 25%, and 30% of the electron mass are used as effective mass approximation for PFET, and NFET, respectively. The measured, and fitted activation energies are in good agreement. The estimated trap energy (E_t) is 0.56eV for the PFET, and 0.50eV for the NFET devices. All parameters of the analytical approximation can be found in the table C.5 of the appendix.

The p^+n S/D diode is simulated by TCAD process, and device simulation to determine the trapping characteristics in the depletion region as basis for transistor simulations. The current due to the Hurkx mechanism, and direct band to band tunneling is simulated. All simulation parameters are listed in section C.4 of the appendix. 8% of the electron mass is used as fitting parameter of the effective mass for the TCAD simulations. The difference, in the approximated effective mass, is due to diverse simplifications that are applied for the analytical model, and the TCAD simulations of the Hurkx current. The differences of the models are explained in detail in section C.5 of the appendix. Table 5.6 compares the results of trap energy and density for the analytical fit, and the simulations. The simulated, and measured activation energies of the PFETs are in good agreement (Fig. 5.14(a)), assuming a mid gap trap.

The silicon oxynitride reference device is compared to the high-k diodes with the lowest leakage: The devices with 3keV Ge implant from the sample set with different gate etches. NFET diode current is increased in the reference devices (see also Fig. C.15(b) of the appendix) due to the higher electric field caused by higher well- and halo implants, and the higher trapping efficiency in the junction (table 5.6). The chosen implant conditions of the p^+n reference lead to a lower S/D leakage current (see also Fig. C.15(a) of the appendix). The metallurgical junction depth of the reference diode is higher, due to an increased boron doping diffusion. This leads to a reduced influence of the halo doping,

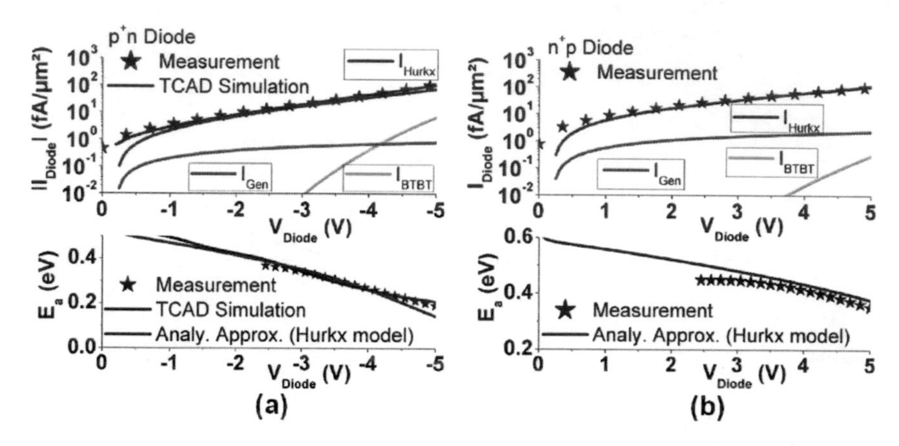

Figure 5.14: Simulated, and measured diode leakage current of thick gate dielectric devices at 85°C are shown in the upper part. The contribution of I_{BTBT}, I_{Hurkx}, and I_{Gen}, calculated by the analytical approximation [43], to the total leakage is presented. The lower part of the graph shows the activation energy change with the bias. Typical curves of the etch sample set for a 10keV germanium implant are shown for:
(a) p^+n diodes.
(b) n^+p diodes.

and a lower electric field. The extracted defect concentration in the junction is similar for the high-k, and the reference device (table 5.6).

Table 5.6: Trapping Characteristics of S/D Diodes***

Device		E_t [eV]	$N_t \cdot \sigma$ [cm^{-1}]	m*	Method
NFET	SiON	0.56	0.12**	0.3	Analytical Fit
	3keV Ge (Etch Group)	0.5	0.03**	0.3	Analytical Fit
PFET	SiON	0.58	0.0048**	0.25	Analytical Fit
	3keV Ge (Etch Group)	0.56	0.0043**	0.25	Analytical Fit
		0.56	0.0012	0.08	TCAD Simulation

**For an assumed thermal velocity of electrons of 230000m/s.
***Median measured curve of the samples is fitted.

5.4.2 Source/Drain Extension Leakage

The source/drain extension leakage (I_{SDE}) occurs from the extension depletion region to the bulk (Fig. 5.13(a)). Figures 5.15(a), and 5.16(a) compare the S/D leakage measured

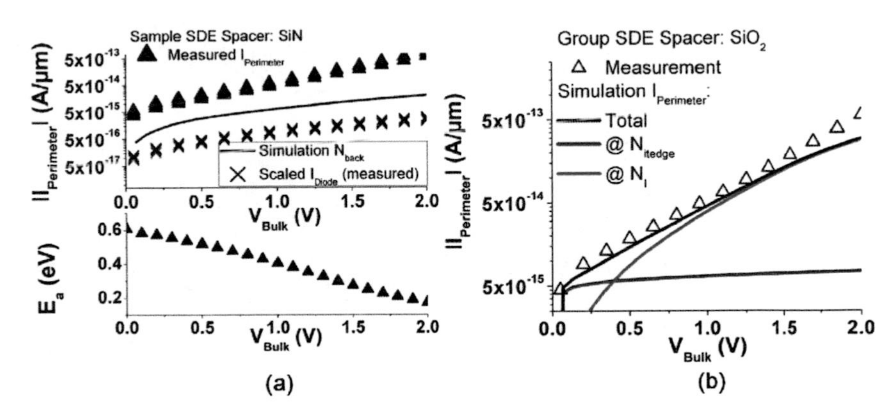

Figure 5.15: Perimeter leakage current of 100nm long thin gate dielectric PFETs from the SDE spacer sample set at 85°C.
(a) The diode leakage is scaled to the source/drain area, and compared to the PFET perimeter current for five dies each in the upper part. The activation energy of the sample with SiN spacer is shown in the lower part of the figure.
(b) The contribution of leakage generated at silicon bulk traps (N_I), and interface traps (N_{itedge}) to the total leakage of a sample with SiO_2 spacer is shown.

at the equivalent diode structures with the PFET perimeter current. The PFET perimeter current consists of the SDE-, and S/D leakage. For comparison, the S/D leakage current is scaled up to the transistor dimensions (equation 3.5). The resulting current is about one decade below the measured perimeter current. Two effects possibly enhance the perimeter current: The increased electric field around the SDE junction, or additional defects. TCAD simulations for the thin gate dielectric PFETs are done using the constant defect density that is extracted in section 5.4.1 in the silicon bulk. The simulated leakage is also significantly smaller than the measured perimeter current. It is concluded that additional defects in the SDE depletion region are enhancing the PFET SDE leakage current.

For the thick gate dielectric NFETs, the scaled source/drain leakage current is in the same order of magnitude like the transistor perimeter leakage (Fig. 5.17). Therefore, it seems that the defect concentration in the region around the NFET extension is similar to the trap density at the vertical source/drain junction (around 80nm-90nm from the silicon surface). There are no additional non-doping implant species used to form the n^+p extension and n^+p source/drain. The amorphization depth of the NFET arsenic SDE implant is about 15nm, and shallow compared to the amorphization depth of the SDE germanium implant of about 23nm that is used in the PFETs. Excess interstitials can be healed out more efficiently [98]. Fewer defects are found in the NFET SDE depletion region by electrical characterization. The SDE leakage of the thin gate dielectric NFETs

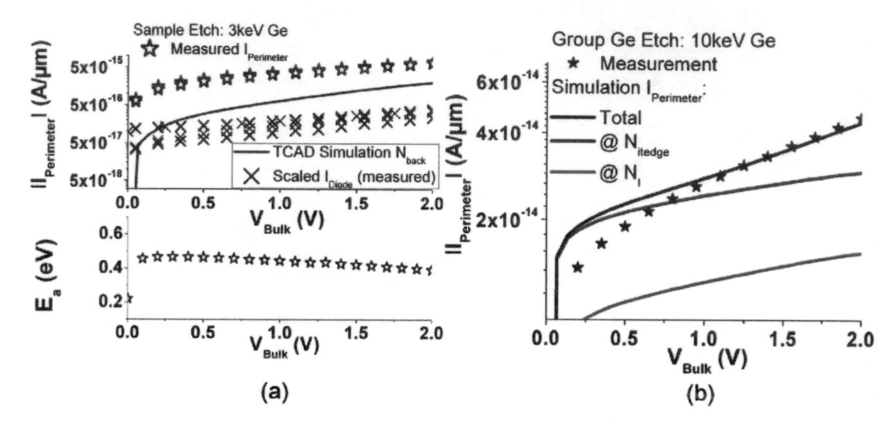

Figure 5.16: Perimeter leakage current of 100nm long thin gate dielectric PFETs from the etch sample set at 85°C.
(a) The diode leakage is scaled to the source/drain area, and compared to the PFET perimeter current for five dies each in the upper part. The activation energy of the sample with a 3keV germanium implant is shown in the lower part of the figure.
(b) The contribution of leakage generated at silicon bulk traps (N_I), and interface traps (N_{itedge}) to the total leakage of a sample with a 10keV germanium implant is shown.

is not measured.

A trap rich region, which is labeled as N_I in figure 5.13(b), around the former amorphous/crystalline (a/c) interface is a possible cause of the increased PFET SDE leakage. It is known, that end of range defects are formed at the a/c interface [95] leaving small clusters of interstitial defects after being healed out. A high interface trap concentration at the gate edge (N_{itedge}) can also enhance the SDE leakage (Fig 5.13(b)).

TCAD device software is used to simulated the SDE leakage of the PFETs with thin gate dielectric, and to determine the main generation center of the current. All model parameters for the TCAD device simulation can be found in section C.4 of the appendix. In the model, the fit parameter of the effective mass is increased for the interface traps compared to the effective mass of the bulk silicon defects. High values of the effective mass fitting parameter for currents arising from interface traps are consistent with literature [51, 99].

The high-k PFETs of the two sample sets differ in their temperature dependence, and slope of the leakage current. The simulated leakage from the charge carrier, generated at N_I, reproduces the measured SDE current of the PFETs from the sample set with different spacers. Below 0.5V also silicon to dielectric interface defects at the gate edge (N_{itedge}) have a significant influence on the SDE current for the transistors of this sample set (Fig. 5.15(b)). A high influence of the interface traps at the gate edge (N_{itedge}) is

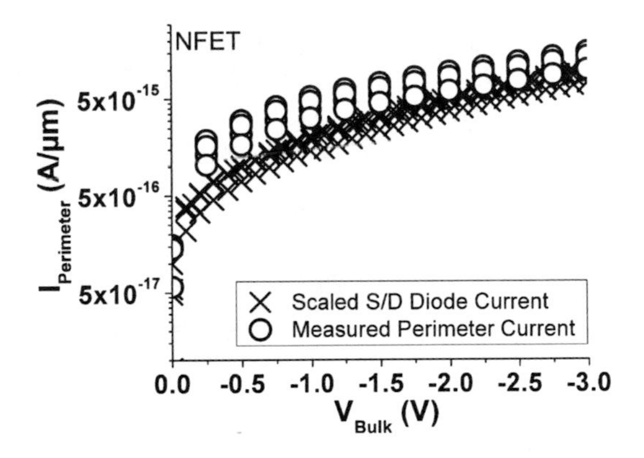

Figure 5.17: Perimeter leakage current of the NFET (L_{Gate}=800nm, CET=5.9nm) from the etch sample set with a 10keV germanium implant at 85°C. Measurements of five dies are shown. The perimeter leakage is in the same order of magnitude as the scaled source/drain leakage.

found in the simulation for the PFETs of the gate etch sample set (Fig. 5.16(b)). The SDE current generated at defects around the former amorphous/crystalline interface (N_I) is about a factor of 2 lower.

The same implantation condition are used for the PFET processing of both sample sets (Fig. 5.18), but the temperature budget and etching chemistry are varied. From the simulation, two orders of magnitude less electrical active defects in the region N_I are found for the etch sample set PFETs due to the improved processing (table 5.7).

Table 5.7: TCAD Trapping Characteristics of PFETs* Determining SDE Current (CET≈2nm)

Group	Device	N_I		N_{itedge}	
		E_t [eV]	$N_t \cdot \sigma$ [cm^{-1}]	E_t [eV]	$N_t \cdot \sigma$ [cm^{-1}]
SDE Spacer	both	0.56	4	0.65	0.00024
Etch	3keV Ge	0.56	0.01	0.65	0.0001
	10keV Ge	0.56	0.04	0.65	0.00165

*Median measured curve of the samples is fitted.

N_I and N_{itedge} are also the main generation centers of the silicon oxynitride reference PFET SDE leakage (see also Fig. C.16 of the appendix). It has a factor five reduced SDE leakage at -1V compared to the high-k PFET with thin gate dielectric of the sample set with different gate etches. No germanium preamorphization implant for gate patterning

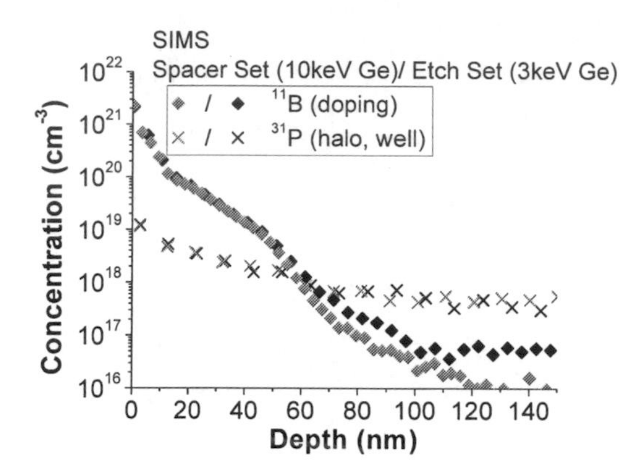

Figure 5.18: SIMS doping profile of the thick gate dielectric PFETs with SiN SDE spacer are shown [85]. The implant conditions of the PFETs are unchanged for both sample sets. Slight variation in the diffusion at the edge of the boron profile are observed, probably due to the changes in etching chemistry, or thermal budget.

is used to process the reference devices. A decreased defect density, N_I and N_{itedge}, for the reference transistor is expected.

5.4.3 Generation Leakage from Channel Region to Bulk

The so called channel leakage ($I_{channel}$) is a generation current which occurs in the depletion region at the gate area (Fig. 5.13(a)). The leakage current is only dominant in long channel devices, and is caused by gate dielectric to silicon interface traps or bulk traps close to the silicon surface (Fig. 2.12). Figure 5.19 shows the measured channel leakage for thin gate dielectric PFET, and NFET. On the test chip, no structures are available to analyze the channel leakage for the thick gate dielectric transistors. The decrease of PFET activation energy with the bias is small (<0.1eV), and also no logarithmic increase in channel leakage with the bias is found, as expected in the generation current model (Fig. 5.20). The NFET channel leakage shows a stronger dependence on the voltage. The NFET channel leakage mechanism is trap assisted tunneling as proposed by Hurkx et al [38]. (section 2.3.3).

The thin gate dielectric PFET current is simulated with the Hurkx model assuming traps at Si/SiO_2 interface (N_{it}). The Hurkx model includes the thermal generation model. Defects with a trap energy of 0.65eV are assumed. Simulation parameters are given in section C.4 of the appendix. A good agreement between the measured, and simulated PFET leakage current is obtained (Fig. 5.19(a)).

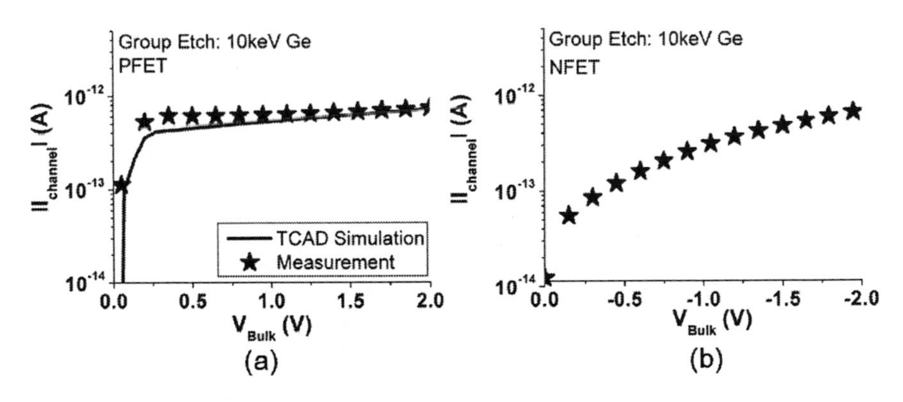

Figure 5.19: Channel leakage of 10μm long devices with thin gate dielectric from the sample set with different gate etches at 85°C is shown.
(a) Comparison of measured, and simulated PFET channel leakage.
(b) Measured NFET channel current is presented.

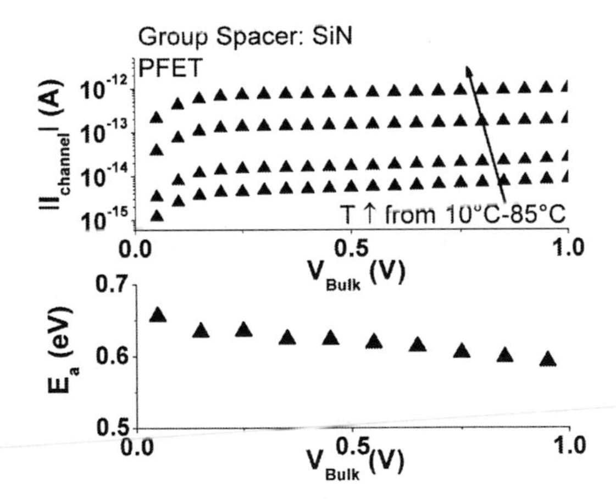

Figure 5.20: Channel leakage of 10μm long thin gate dielectric devices from the sample set with different SDE spacers. The temperature dependence of a typical device is presented in the upper part of the graph. The activation energy depending on the bias is shown in the lower part of the figure.

Table 5.8 gives an overview of the measured channel currents, and interface defect densities for the thin gate dielectric devices. For the PFETs, the channel leakage increases with the interface defect density. The NFET channel leakage of the silicon oxynitride reference device is increased compared to the high-k device. This can be explained by the higher V_{th} doping concentration in the reference NFET.

Table 5.8: Trapping characteristics of the Transistor* Channel Leakage (CET≈2nm)

Device	Group	$I_{channel}$	N_{it} (CP)	$N_{it} \cdot \sigma$ (TCAD)
		[A/μm^2]	[cm^{-2}]	[cm^{-1}]
PFET	SDE Spacer	$1.2{\cdot}10^{-14}$	$7.2{\cdot}10^{10}$	$1.8{\cdot}10^{-5}$
	Etch	$6.3{\cdot}10^{-15}$	$4.8{\cdot}10^{10}$	$8.8{\cdot}10^{-6}$
	SiON Reference	$3.0{\cdot}10^{-15}$	$1.3{\cdot}10^{10}$	-
NFET	Etch	$6.2{\cdot}10^{-15}$	$4.0{\cdot}10^{9}$	-
	SiON Reference	$3.9{\cdot}10^{-14}$	$4.7{\cdot}10^{9}$	-

* Channel leakage measured at ±1V. The average $I_{channel}$, and interface trap density, determined by CP, of five dies is given. The median leakage curve is fitted using TCAD simulation.

5.4.4 Gate Induced Drain Leakage

The gate induced drain leakage (GIDL) is especially sensitive to defects at the gate edge (Fig. 5.21). Due to the source/drain implantation process steps, the defect density at the gate edge is expected to be higher than in the channel area [72].

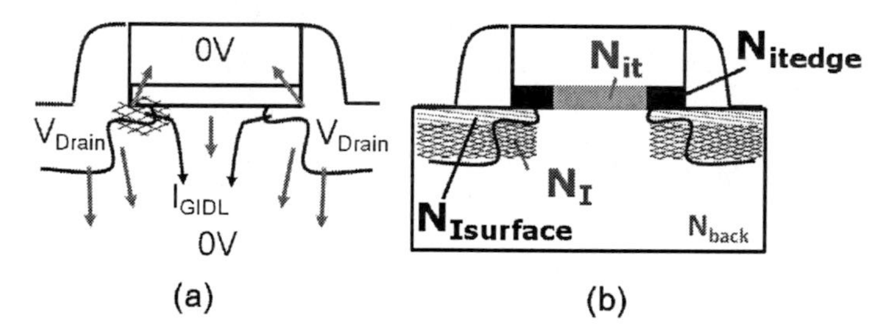

Figure 5.21: Schematic sketch of the traps responsible for generating GIDL.
(a) GIDL current (black) occurrence in the MOSFET. The other leakage paths under the chosen measurement conditions are marked in gray.
(b) Trap distribution of the transistor assumed in the simulations. GIDL is mainly caused by N_{itedge} and $N_{Isurface}$.

The GIDL current occurs at a deep depletion that forms in the overlap region of the S/D extension underneath the gate (section 2.3). There are two components influencing

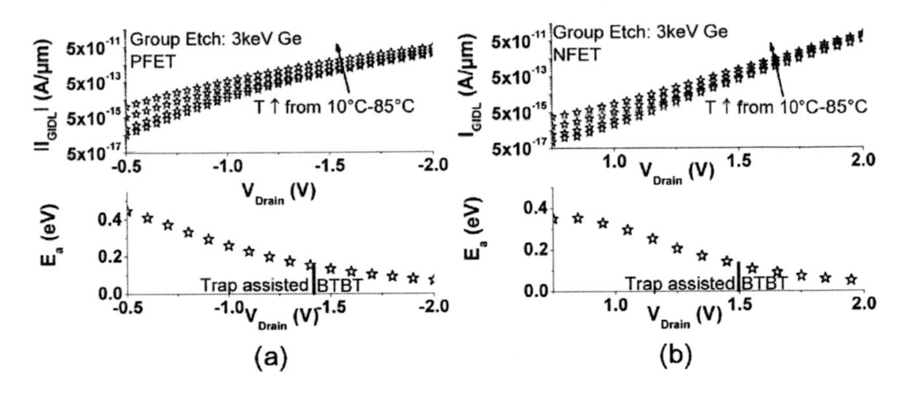

Figure 5.22: The measured temperature dependence of the GIDL current for thin gate dielectric transistors from the etch set is shown in the upper part of the figure. The activation energy is plotted in the lower part of the figure. The main leakage mechanisms is changing from a trap assisted tunneling to a band to band tunneling when the bias increases.
(a) PFET
(b) NFET

the GIDL. Band to band tunneling is nearly not temperature dependent, with an activation energy below 0.15eV [32]. Trap assisted tunneling occurs when charge carriers are thermally activated, and tunnel through the remaining barrier [51]. This phenomenon is found for the bulk traps close to the surface ($N_{Isurface}$), as well as for silicon to gate dielectric interface traps at the gate edge (N_{itedge}) (Fig. 5.21(b)).

The temperature dependences of the thin gate dielectric transistor GIDL is shown in figure 5.22. Similar plots for the thick gate dielectric transistors can be found in figure C.14 of the appendix. The activation energy decreases with the voltage for a thermally assisted tunneling mechanism via defects after Hurkx et al. [38] down to 0.15eV. Once the activation energy reduces below 0.15eV, the band to band tunneling mechanism dominates the leakage current.

The GIDL current is simulated for the 100nm long thin gate dielectric PFETs. The simulated- and measured leakage current increase with voltage are in good agreement (Fig. 5.23). A one level mid gap trap is assumed for the bulk silicon trap simulations. For the interface defects, a trap energy of 0.65eV is assumed. The simulation parameters are listed in section C.4 of the appendix.

The leakage of the thin gate dielectric PFETs is mainly generated at interface traps (Fig. 5.23). The trapping characteristics of the simulated GIDL for the PFETs with thin gate dielectric are summarized in table 5.9. For thick gate dielectric PFETs $N_{Isurface}$, and N_{itedge} contribute evenly to the overall GIDL (section 4.4.4). The GIDL current of the

Table 5.9: TCAD Trapping Characteristics of PFETs* (CET≈2nm) Determining GIDL

Group	Device	$N_{Isurface}$		N_{itedge}	
		E_t [eV]	$N_t \cdot \sigma$ [cm^{-1}]	E_t [eV]	$N_t \cdot \sigma$ [cm^{-1}]
SDE Spacer	both	0.56	0.04	0.65	0.00024
Etch	3keV Ge	0.56	0.0001	0.65	0.0001
	10keV Ge	0.56	0.0004	0.65	0.00165

* The median leakage curve is fitted using TCAD simulation.

NFET devices is relatively insensitive to the interface traps at the gate edge (section 2.3.2) [50]. The trap assisted NFET GIDL current is more dependent on silicon bulk traps close to the surface ($N_{Isurface}$).

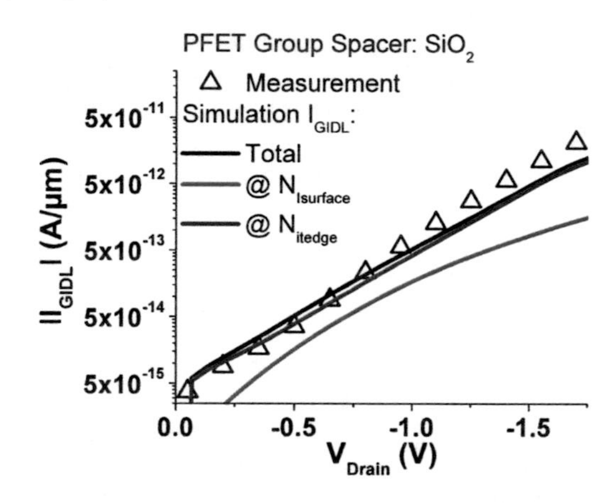

Figure 5.23: Simulated, and measured GIDL current for a 100nm long PFET with thin gate dielectric at 85°C is shown. The current of a typical transistor with SiO_2 SDE spacer is shown in the figure. The GIDL current generated at the interface defects (N_{itedge}), and at the bulk silicon defects ($N_{Isurface}$) is presented.

Nearly no temperature dependence is found for the GIDL measured at the silicon oxynitride reference NFETs (see also Fig. C.17 of the appendix). Due to the higher doping concentration, a higher band to band tunneling leakage occurs. The GIDL of the silicon oxynitride, and high-k PFET show the same temperature-, and voltage dependence. The lower GIDL current of the thin reference PFETs is probably caused by a reduced damage (table 5.5) because no high-k preamorphization implant is used.

5.4.5 Gate Leakage

The high-k material is introduced in the gate stack to reduce the gate leakage (I_{Gate}) by increasing the physical dielectric thickness without raising the CET [12]. Figure 5.24 shows the NFET, and PFET gate leakage in dependence on the CET for the high-k stacks, and the silicon oxynitride reference sample. The gate leakage differences due to gate edge processing will be discussed in detail in the last two sections of this chapter.

For the high-k transistors with a thin gate dielectric, a clear reduction in gate leakage by about a factor of 20 compared to the reference is found. The gate leakage of the reference sample is within the noise level of the measurements for the thick gate dielectric transistors (not shown). For the thick high-k stack, the gate leakage is increased (Fig. 5.24). This unexpected gate leakage increase points towards a non-optimized integration. However, gate leakage is only a small contribution to the overall off-current of the MOSFETs with thick gate dielectric (see also section C.2 of the appendix). The PFETs with different spacers have a slightly reduced gate leakage probably due to the changes in etching chemistry, and thermal budget.

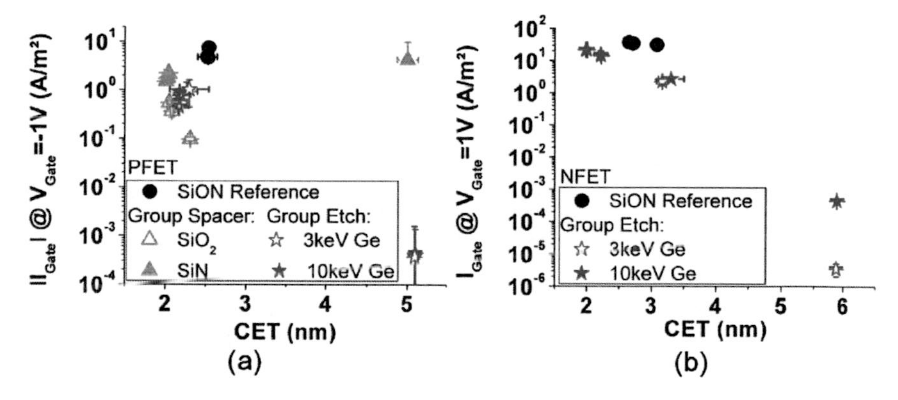

Figure 5.24: Measured gate leakage depending on CET at ±1V inversion bias for devices with different channel length. The leakage current at 85°C of the high-k gate stack is compared to a silicon oxynitride reference for:
(a) PFETs.
(b) NFETs.

The activation energy, calculated from the temperature dependent current voltage curves of the gate current, is small below 0.15eV (see also Fig. C.20 of the appendix). Therefore, direct band to band tunneling causes the leakage current [32]. Mainly two conduction mechanisms are described in literature: Fowler Nordheim tunneling through a triangular barrier, and direct tunneling through a trapezoid barrier (section 2.1) [26].

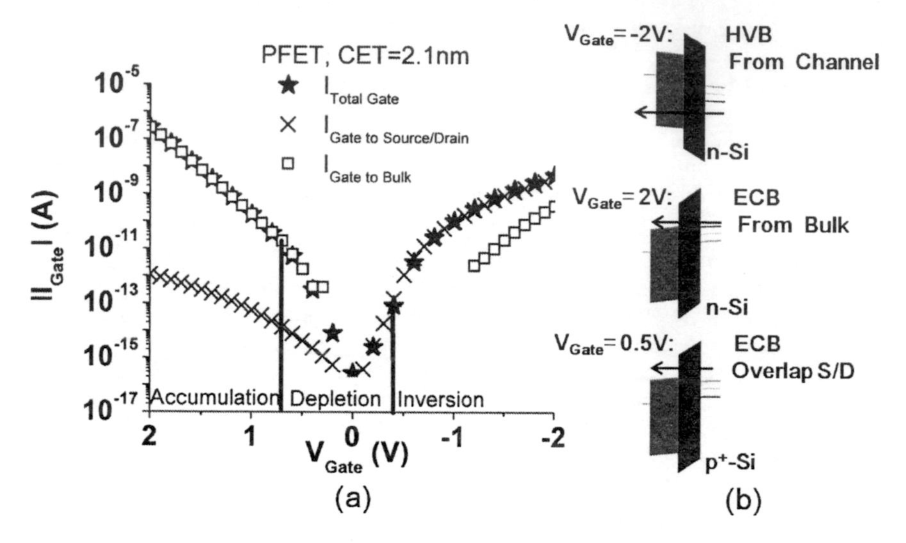

Figure 5.25: The gate leakage of a 0.8μm long PFET from the etch sample set (10keV germanium) at 85°C is presented.
(a) Contribution of gate to source/drain, and gate to bulk leakage on overall gate current depending on the voltage are shown for the thin gate dielectric device.
(b) Sketch of the leakage flow due to hole valance band tunneling (HVB) at V_{Gate}= -2V, and electron conduction band tunneling (ECB) at V_{Gate}=2V is presented. The source/drain overlap current dominates the depletion regime (V_{Gate}=0.5V) [35].

Figure 5.25 presents the dependence of the gate leakage on the voltage for the 800nm long PFET with thin gate dielectric. The same investigation is done for the 800nm long NFET with thin-, and thick gate dielectric (section C.8 of the appendix). The gate current depending on the bias can be divided in three regions (Fig. 2.2(a)). For the thin gate dielectric PFET, an area dependent electron conduction band tunneling current (ECB) to the bulk is observed when the gate is biased to accumulation. In the depletion regime, an ECB leakage over the source/drain overlap occurs. In inversion, an area dependent hole valence band tunneling current (HVB) to the channel arises [35]. The NFET shows an area dependent electron valence band current (EVB) under accumulation, an ECB tunneling over the source/drain overlap in the depletion regime, and an area dependent ECB tunneling from the gate to the channel in inversion [35].

The area dependent Fowler Nordheim leakage, and direct tunneling current of the devices with thin gate dielectric are simulated with the BSIM4 model [74], and the ngspice program [55]. The simulation parameters of the model are given in section C.9 of the appendix. The model is fitted to the leakage current measured at the 10μm long device. In the next step, the results are compared to shorter devices leaving only the CET as free

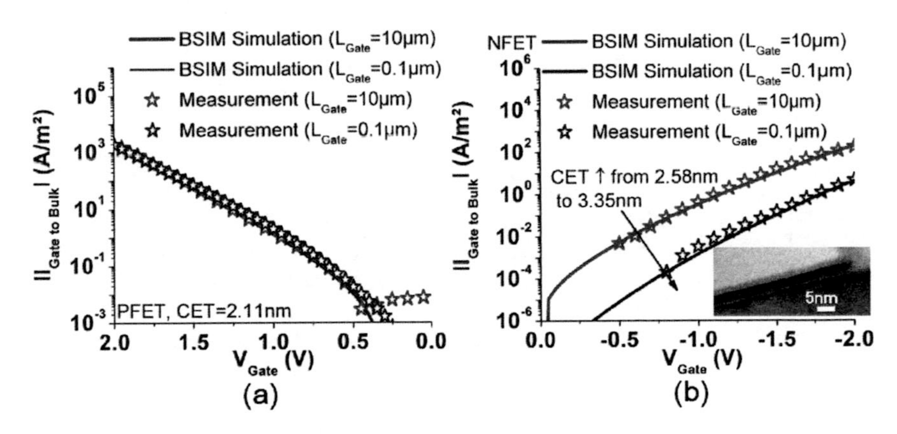

Figure 5.26: Simulated (BSIM), and measured gate to bulk leakage of thin gate dielectric devices from the sample set with different etches (3keV Ge) at 85°C. The median measured curve is presented.
(a) No gate length dependence of the area scaled leakage is observed for the PFET.
(b) The NFET shows a decrease in area scaled leakage with lower channel length. An interfacial layer between the high-k and the polysilicon is observed in the TEM analysis in the inset [115].

fit parameter.

Figure 5.26 gives an example of the fitted leakage current for a 10μm and 100nm long PFET, and NFET with silicon nitride extension spacer. For the PFET, gate leakage scales with the area, and is independent of the channel length. For the NFET, a higher CET has to be assumed to fit the short channel gate leakage. A 1.2 nm CET increase is also found by capacitance voltage measurements when the NFET channel length is decreased from 10μm to 100nm.

In TEM analysis, an interfacial layer at the NFET high-k polysilicon interface is observed. But no clear edge effects are observed, explaining the increase of CET with reduced channel length. One possible explanation is a nitridation at the polysilicon to gate dielectric interface from the gate edge due to the silicon nitride spacer.

5.4.6 Subthreshold Leakage

The subthreshold leakage (I_{SubVth}) is a current between source, and drain when the gate bias is below V_{th} (Fig. 5.10(b)). The subthreshold current is mainly dominated by the diffusion current [3]. It is enhanced by the drain induced barrier lowering effect (DIBL), for devices with short gate length (section 2.2).

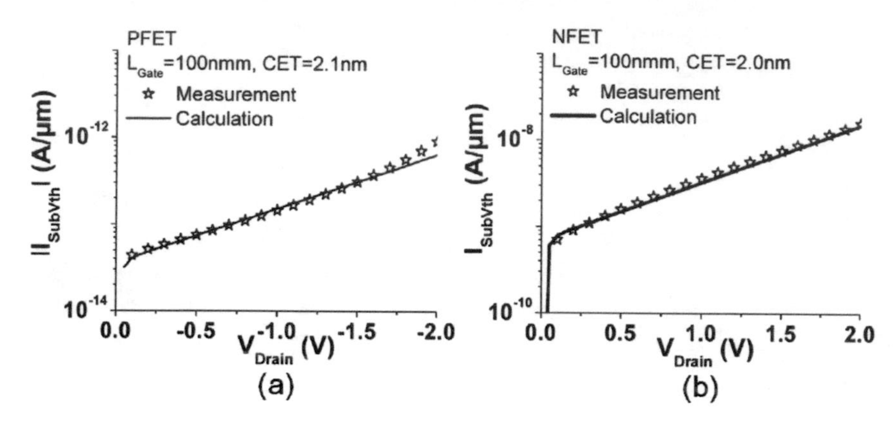

Figure 5.27: Measured, and calculated subthreshold current for 100nm long thin gate dielectric transistors from the etch sample set (10keV germanium).
(a) PFET subthreshold current is approximated using a fitted DIBL parameter of 0.055, and a V_{th} of -0.67V.
(a) NFET subthreshold current is calculated using a fitted DIBL parameter of 0.065, and a V_{th} of 0.45V.

The subthreshold current is calculated using equation (4.1) [100]. All fitting parameters are listed in table C.11 of the appendix. The DIBL parameter (λ) depends on the type of the gate dielectric, the gate length, and the V_{th}- and halo doping (body factor m) (equation 2.5). It determines the slope of the subthreshold current with the drain voltage. The calculated, and measured currents are in good agreement (Fig. 5.27).

Table 5.10: Parameters of Transistor Subthreshold Current (CET≈2nm)

PFET		V_{th} (V)*	λ**	m*
SiON		0.57	0.04	1.5
Group Etch	3keV Ge	0.67	0.055	1.28
Group Etch	10keV Ge	0.66	0.055	1.28
Group Spacer	SiN	0.63	0.045	1.34
Group Spacer	SiO_2	0.62	0.075	1.44
NFET		V_{th} (V)*	λ**	m*
SiON		0.52	0.03	1.36
Group Etch	3keV Ge	0.45	0.065	1.38
Group Etch	10keV Ge	0.44	0.065	1.40

* Extracted from the transfer curves of 100nm long devices at V_{Drain}=±0.05V.
** Extracted from the fit of $I_{SubVth}V_{Drain}$ of 100nm long devices.

The main parameters, that effect the subthreshold current, are summarized in table 5.10. The DIBL parameter is increased for the high-k gate stacks compared to the reference silicon oxynitride sample mainly due to the reduced halo, and V_{th} doping (see also section C.1 of the appendix). In comparing the high-k PFET with the silicon oxynitride reference device, a reduced influence of the subthreshold current on the overall off-current of the high-k PFET is found (table 5.5). That is caused by the higher threshold voltage due to the metal gate stack of the high-k PFET.

5.4.7 Conclusion

In this section, the various leakage currents under off conditions for the high-k devices have been investigated. The leakages are compared to the current measurements of a standard transistors with silicon oxynitride gate stack based on the 65nm technology.

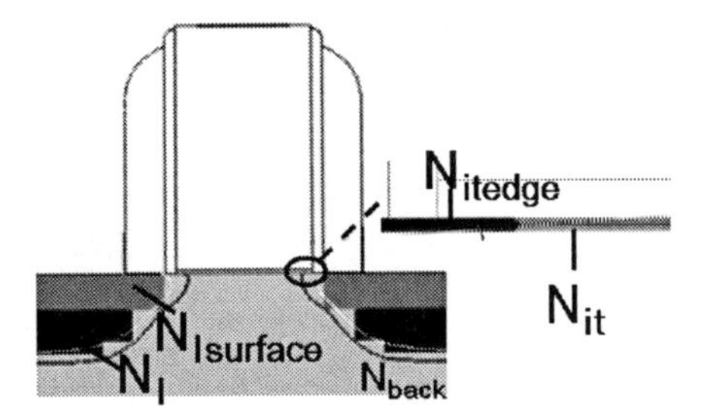

Figure 5.28: Placement of the continuous defect bands in the simulations presented for a PFET with a CET of 2.1nm and L_{Gate}=100nm.

The gate leakage is reduced by a factor of about 20 for high-k devices with thin gate dielectric compared to a silicon oxynitride reference stack (Fig 5.24). The NFET shows a decrease in gate leakage per area with shorter channel length (Fig. 5.26). From the BSIM model, a 0.75nm CET increase between 10μm, and 100nm channel length is consistent with the gate leakage decrease.

From transmission electron microscopy analysis of the NFET, an interfacial layer between the polysilicon, and the high-k layer is observed. No visible edge effects are found. A possible explanation of the NFET short channel CET increase is nitridation of the interface at the gate edge from the spacer. The high-k devices with thick gate dielectric show an unexpectedly high gate leakage current.

For the PFETs with thin gate dielectric, a TCAD Sentaurus simulation model with different defect regions is developed, to explain source/drain-, channel-, source/drain extension-, and GIDL current (Fig. 5.28). It is based on a generation current, enhanced by trap assisted tunneling (Hurkx [38]). Additionally, direct band to band tunneling is included in the model. The model explains different leakage current paths, and the current voltage dependence. The Frenkel Poole effect, avalanche multiplication, and the saturation current are not needed to reproduce the leakage current characteristics. The TCAD model is developed to analyze the effect of the process changes on the defect density at the gate edge in sections 5.5, and 5.6.

The trap distribution in the PFET model is discrete, using trap bands with constant defect density (Fig. 5.28). The accurate extraction of the real values of the trap density is ambitious because of two effects. First, the capture cross section is not known. Second, the simulation reacts sensitively to the fit value of the trap density, and the placement of the discrete bands. The advantage of the model is that the leakage currents, and defect densities of two similar devices can be compared. In addition, a rough comparison of the trap concentration in different defect regions of one transistor is possible.

Table 5.11 gives an overview of the extracted defect concentrations for the PFETs from the two sample sets. The capture cross section is set to be constant for the bulk silicon, and silicon to gate dielectric interface defects in this model, to compare absolute numbers of the defect density.

The perimeter current of the high-k-, and the silicon oxynitride reference PFETs is increased compared to the source/drain diode leakage when the bulk is biased. That is caused by an additional current generation from defects in the source/drain extension depletion region (N_I). The increased trap density is probably due to interstitial defects from the germanium amorphization implantation. Also, a high amount of interface defects at the gate edge (N_{itedge}) leads to the increased leakage. For the NFETs, the perimeter current is mainly source/drain diode leakage. A shallow arsenic doping implant amorphizes the NFET source drain extension. Interstitial defects close to the silicon surface can be healed out more efficiently [98].

Table 5.11: Simulated Concentration of Defects of High-k PFETs (CET≈2)

Trap Region	N_t [cm^{-3}]			σ [cm^2]	E_t [eV]
	Group Spacer	Group Etch			
		10keV	3keV		
N_{back}	$9.2 \cdot 10^{12}$	$5.5 \cdot 10^{12}$	$1.2 \cdot 10^{12}$	$1 \cdot 10^{-15}$	0.55
N_I	$4.0 \cdot 10^{15}$	$4.0 \cdot 10^{13}$	$1.0 \cdot 10^{13}$		
$N_{Isurface}$	$4.0 \cdot 10^{13}$	$4.0 \cdot 10^{11}$	$1.0 \cdot 10^{11}$		
N_{it}	$1.8 \cdot 10^{10}$	$8.8 \cdot 10^{9}$		$1 \cdot 10^{-15}$	0.65eV
N_{itedge}	$2.4 \cdot 10^{11}$	$1.7 \cdot 10^{12}$	$1.0 \cdot 10^{11}$		

The PFET leakage from the channel region (channel leakage) is created at silicon to gate dielectric interface traps (N_{it}). The silicon oxynitride reference transistor has a decreased interface trap density, and reduced channel leakage. The high-k PFET uses an aluminum oxide layer to tune the threshold voltage, and a titanium nitride gate electrode. This gate stack leads to a higher interface defect density. Charge pumping measurements show that the high-k-, and reference NFET with thin gate dielectric have a similar interface defect density. The NFET silicon oxynitride reference has a higher V_{th} doping, and increased channel leakage.

The GIDL current mechanism of high-k NFETs, and PFETs changes from trap assisted tunneling to band to band tunneling when the bias is increased. Simulations confirm the expected increase in trap density at the gate edge (N_{itedge}) for the PFETs (table 5.11). In case of the thick gate dielectric PFETs, both interface traps at the gate edge (N_{itedge}), and silicon bulk traps close to the interface ($N_{Isurface}$) contribute to the overall GIDL. The silicon oxynitride NFET GIDL is caused by band to band tunneling only. In contrast to the reference, charge generation at the $N_{Isurface}$ increases the high-k NFET GIDL (Fig. 5.22(b)). The additional germanium implant for high-k preamorphization probably leads to a higher defect density, and the lower doping implants to a reduced electric fields in the high-k transistors.

The subthreshold leakage is the major off-current for all short channel NFETs when the gate is biased to 0V, and the drain voltage is ramped while the source voltage is kept constant.

5.5 Comparison of Silicon Oxide and Silicon Nitride Extension Spacer

By default silicon oxide extension spacers are used in the Qimonda process flow. Silicon oxide spacers are relatively cheap, and easy to integrate. Silicon nitride extension spacers are reported to have advanced on-current characteristics [120]. The higher permittivity value of silicon nitride directly effects the electric field at the gate edge [121]. Silicon nitride is also known to have a higher defect density [122, 123]. The influence of the dielectric material on the leakage currents is investigated in this section.

5.5.1 Description of Experiment

In the experiment, PFETs with silicon oxide - and silicon nitride SDE spacer are compared. TEM measurements reveal that the silicon oxide SDE spacer leads to an increasing interface oxide thickness at the gate edge (Fig. 5.29). The undesired phenomenon is similar to the intentional growth of the so called birds beak [124]. The oxygen from the spacer leads to a regrowth of the 1nm thick silicon oxide interlayer.

5.5.2 Electrical Characterization of Transistor Performance

The most important basic electrical parameters depending on the SDE spacer material are summarized in table 5.12 for the thin gate dielectric PFETs. A comparison of the long

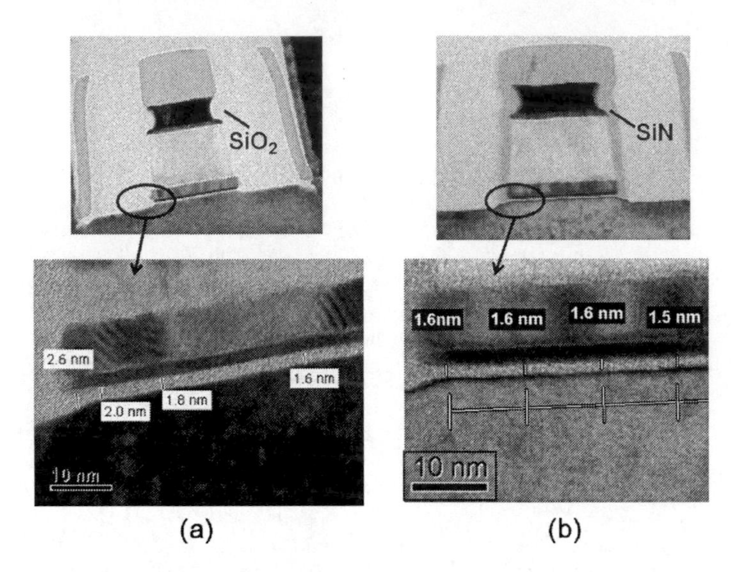

Figure 5.29: TEM analysis of the PFETs is presented.
(a) PFET with SiO_2 SDE spacer [125]
(b) PFET with SiN SDE spacer [115]

channel thin-, and thick gate dielectric devices can be found in tables C.12, C.13 of the appendix. In comparing 100nm long devices to 10μm long transistors, a CET increase by 0.3nm is found by split capacitance voltage measurement of PFETs with SiO_2 SDE spacer. No such CET increase is observed for the PFETs with SiN SDE spacer. The results are explained by the bird beaks phenomenon that is observed in the TEM measurement (Fig. 5.29). The overlap length is reduced by about 1.5nm for the PFETs with SiN SDE spacer. That could be due to the variation in the SDE spacer thickness for the two processes.

5.5.3 Leakage Current Dependence on the Spacer Dielectric

The changes of the leakage currents due to the variation of the SDE spacer material will be analyzed in this part of the section. Variations of the electric field, and of the defect density at the silicon to spacer interface will be taken into account.

Source/Drain Extension Leakage

Source/drain extension leakage is not significantly different if the SDE spacer material is changed (Fig. 5.30). The statistical variation of the PFET SDE current is with a factor of 4 at a bias of -1V. The slope of the measured leakage with the voltage is slightly varying.

Table 5.12: Basic Characteristics of PFETs with Different SDE Spacers (CET≈2nm)

	L_{Gate}	SiO_2 Spacer	SiN Spacer
CET (nm)	0.1μm	2.31±0.06	2.03±0.04
	10μm	2.05±0.02	2.01±0.02
L_{ov} (nm)	0.1μm	5.1±0.4	3.3±0.2
	10μm	4.5±0.6	2.9±0.6
C_{Diode} (fF/μm²) V_{Diode}=-1V (CV*)	/	1.61±0.3	1.57±0.02
V_{th} (V) V_{Drain}=-0.05V (CV**)	0.1μm	-0.61±0.07	-0.6±0.05
	10μm	-0.53±0.08	-0.52±0.13
V_{th} (V) V_{Drain}=-0.05V (IV)	0.1μm	-0.62±0.09	-0.63±0.01
	10μm	-0.59±0.01	-0.59±0.02
V_{th} (V) V_{Drain}=-1V (IV)	0.1μm	-0.56±0.05	-0.55±0.07
V_{fb} (V) (CV**)	0.1μm	0.56±0.07	0.55±0.07
	10μm	0.48±0.13	0.51±0.07
I_{on} (μA/μm)	0.1μm	59.7±14.0	49.3±3.8
	10μm	0.82±0.05	0.87±0.08
S_{Vth} (mV/dec) V_{Drain}=-0.05V	0.1μm	99.2±5.6	94.7±4.5
	10μm	80.0±1.8	79.4±1.4
λ (IV***)	0.1μm	0.058±0.016	0.058±0.004

All values are given for measurements on 5 dies (wafer center). All PFETs have a 10keV Ge preamorphization of the high-k.
* Capacitance voltage measurements are done on a diode structure with SDE and S/D implants.
** V_{fb} and V_{th} are taken from the point of maximum, and minimum slope of the CV characteristic.
*** DIBL parameter is determined after the V_{th} shift of $I_{Drain}V_{Gate}$ as described in section 2.2.

The simulations predict the main SDE current generation at the interstitial rich region around the former amorphous/crystalline interface (N_I, Fig. 5.15(b)). No influence of the SDE spacer material on the current generated at N_I is expected from the simulations. The measured decrease in overlap length of the PFETs with silicon nitride extension spacer is reproduced in the simulations by a 3nm thicker SDE spacer. A slight increase in the simulated SDE current for the PFET with SiN spacer due to a reduction in overlap length is found (Fig 5.31 upper curve). The increase in current is mainly caused by the higher number of defects spreading in the SDE depletion region.

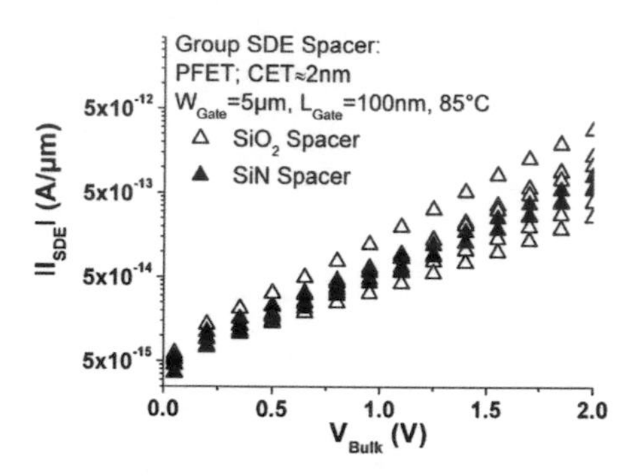

Figure 5.30: Measured SDE leakage of the 100nm long PFETs with thin gate dielectric. Five dies are measured of each sample of the spacer set with SiO_2-, or SiN SDE spacer at 85°C.

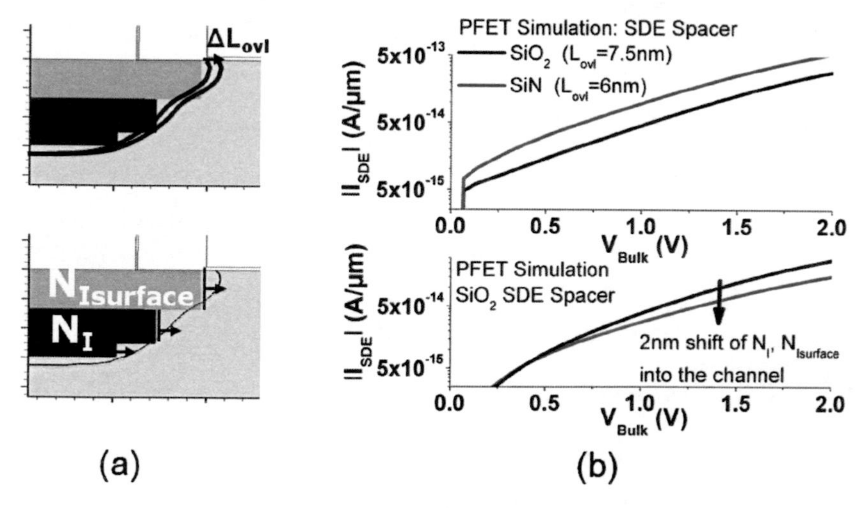

Figure 5.31: The upper curve shows the simulated variation of the SDE leakage with the overlap length of PFETs with different SDE spacers. The lower curves represents the effect of changes in defect placement on the current generated from bulk silicon traps. A constant total amount of traps is assumed for both curves in the lower graph by adjusting the defect density.
(a) Schematics of the simulated variations of L_{ov}, and the defect placement (N_I, $N_{Isurface}$) are shown.
(b) Simulated current voltage plots of thin gate dielectric PFETs at 85°C are presented.

However, the predicted change in SDE current by a factor of 2.2 is within the variation of the measured curves. The measured differences in the slope of the leakage is reproduced in the simulation if the defect band N_I diffuses a few nanometer more or less towards the middle of the channel (Fig 5.31 lower curve). In conclusion, the simulated changes for the samples with silicon oxide, or -nitride spacer are within the statistical deviation of the measurement. The variation in defect density, and -placement is higher than the effect of the variation of overlap length between the samples.

Gate Induced Drain Leakage

An increase in trap assisted GIDL is found for the PFETs with the silicon nitride SDE spacer (Fig 5.32). The increase is in average a factor of 3 in the trap assisted GIDL regime at -1V for the thin gate dielectric PFETs. For the thick gate dielectric transistors the leakage current increases in average by a factor of 1.5 in the trap assisted GIDL regime at -1V. Different factors effect the GIDL: The reduction of oxide thickness at the gate edge, due to the suppression of the birds beak phenomenon, the higher permittivity value of the silicon nitride spacer, and the reduced overlap length alter the electric field. Additionally, an increase of silicon to spacer interface defect density for the silicon nitride SDE spacer possibly raises the leakage.

TCAD simulations for the PFETs with thin gate dielectric are done to separate the factors influencing the trap assisted GIDL. Band to band tunneling dominates the GIDL current at -1.4V to -2V, and -1.75V to -3.1V for thin, and thick gate dielectric PFETs, respectively. Band to band tunneling is not analyzed in detail. Simulations show that the GIDL current is generated mainly by the dielectric to silicon interface traps (Fig. 5.23).

The higher permittivity value of the SDE spacer increases the slope of the interface trap assisted GIDL between -0.5V and -2V (Fig. 5.33(a)). The GIDL is increased by a factor of 1.7 at -1V. Figure 5.34(a) shows that the maximum of the GIDL generation shifts more to the gate edge at higher voltages where the SDE spacer material has more influence.

Also, the interface trap assisted GIDL is increased by the decrease in CET at the gate edge by 0.3nm (Fig. 5.33(a)). The CET decrease increases the GIDL by a factor of 1.8 at -1V. The electric field at the gate edge is increased from 2.7MV/cm to 2.9MV/cm at -1V with the lower thickness (Fig. 5.34(b)). The reduction in overlap length slightly reduces the slope of the interface trap assisted current (Fig. 5.33(a)). The GIDL reduces by a factor of 1.3 at -1V for a 1.5nm decrease in overlap length for the PFET with silicon nitride spacer.

The simulations reproduce the minimal interface trap assisted GIDL current increase due to different spacer materials without any changes in electrical active defects (Fig. 5.33(b)). But, the GIDL simulations are very sensitive to CET, and overlap length variations. The accuracy of the split capacitance voltage measurements is about 0.1nm for the CET, and 1nm for the overlap length. High silicon to spacer interface defect densities for silicon nitride are reported in literature [122, 123], and can not be ruled out in the simulations.

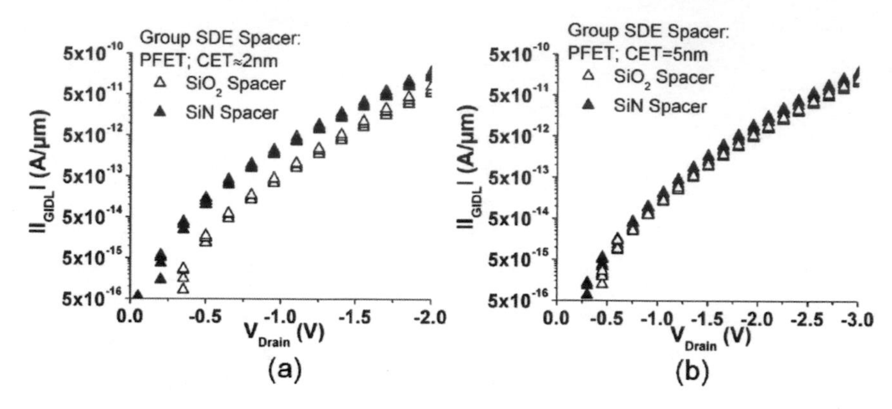

Figure 5.32: Gate induced drain leakage for transistors with different SDE spacer materials at 85°C, and at a gate bias of 0V is presented. The figure presents the measured GIDL of five dies for:
(a) PFETs with thin gate dielectric.
(b) PFETs with thick gate dielectric.

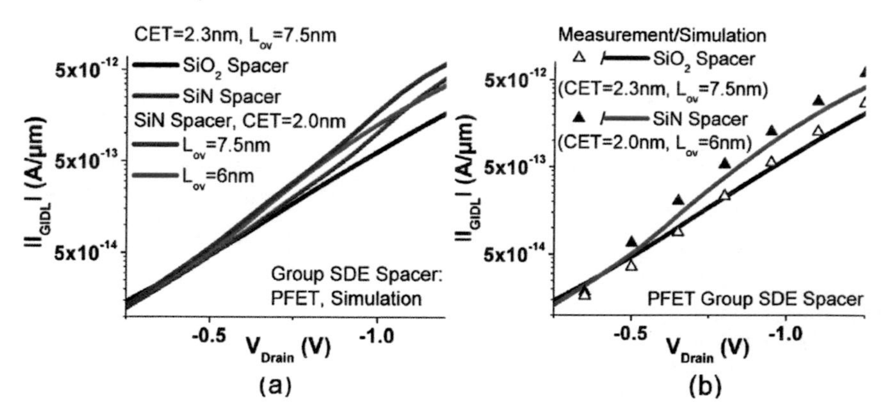

Figure 5.33: Measured, and simulated GIDL for PFETs with thin gate dielectric at at 85°C is presented.
(a) Simulated GIDL current depending on SDE spacer material, CET, and L_{ovl} is shown.
(b) Comparison of minimal measured, and simulated GIDL difference for PFETs depending on the SDE spacer material without changes in defect density is presented.

The difference in GIDL current due to the change in SDE spacer material is smaller for the thick gate dielectric PFETs (Fig 5.32(b)). That may be caused by the fact that bulk

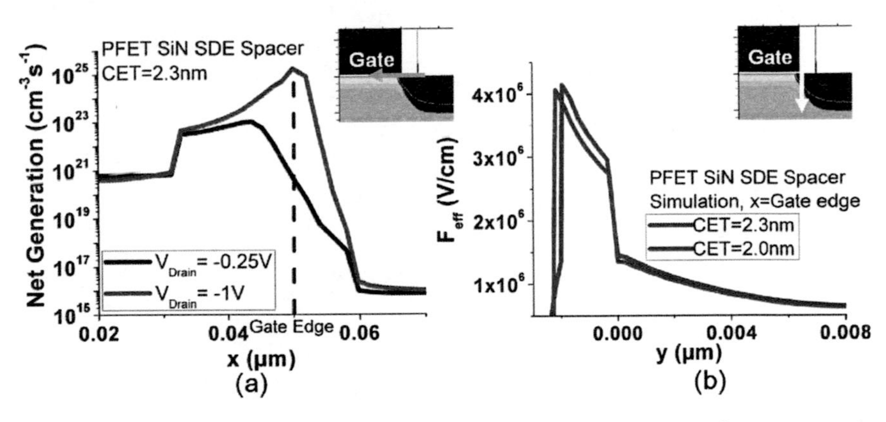

Figure 5.34: Simulation of the PFET properties at the gate edge at 85°C are presented:
(a) Simulated net generation at the silicon to dielectric interface traps for different biases
(b) Simulated vertical electric field for PFETs with SiN SDE spacer at different CET

silicon $N_{Isurface}$ defects contribute significantly to the overall GIDL (section 5.4.4). The current generated at these defects is not significantly effected by the spacer.

Gate Leakage

The gate leakage per area depending on the CET is shown in figure 5.35(a). The PFETs with silicon oxide SDE spacer show a significant decrease in gate leakage for shorter channel length, which is related to unintended bird beaks (Fig. 5.29). The edge effect leads to a higher average CET of short channel MOSFETs with thin gate dielectric. From the TEM analysis, a rapid decrease of the oxide regrowth towards the middle of the channel is extracted (Fig. 5.36(a)). Therefore, the average CET of the long channel devices is not effected. Due to the unintended bird beaks, the gate to source/drain overlap leakage current is reduced for the devices with silicon oxide SDE spacer. With split capacitance voltage measurement, an increase in average CET of 0.3nm is measured for the short channel PFETs with silicon oxide spacer. From the simulation of the gate leakage current under inversion, an average CET increase of 0.1nm is extracted (Fig. 5.36(b)).

The thick gate dielectric devices show a high leakage, especially for the PFETs with SiN SDE spacer (Fig. 5.35(b)). The processing of the sidewalls is one possible cause for the increased gate current. The leakage of the transistors with a thick dielectric stack is improved by production changes in etching, and thermal budget (see also Fig. C.21 of the appendix).

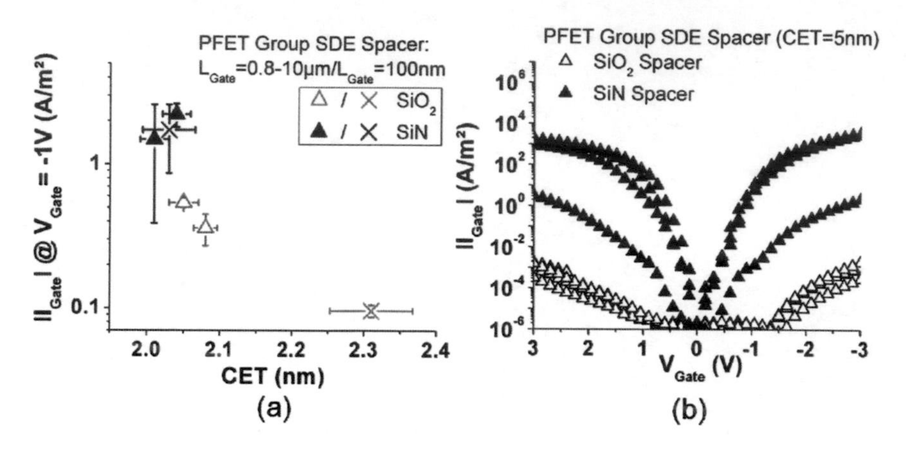

Figure 5.35: (a) Measured gate leakage current depending on CET for PFETs with the thin gate dielectric stack at 85°C is shown.
(b) Gate leakage of 800nm long PFETs with thick gate dielectric depending on the bias at 85°C is presented.

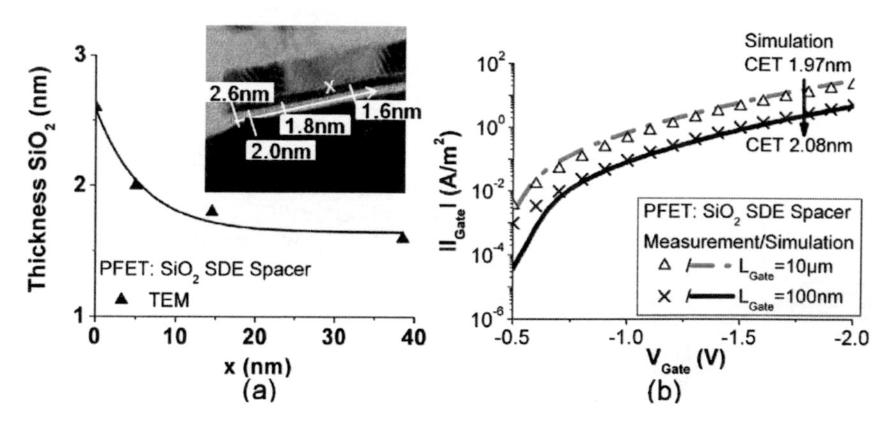

Figure 5.36: (a) Measured SiO_2 interlayer thickness, determined by TEM analysis [125], depending on the placement between the gate edge and - middle. A rapid decay of the thickness is observed for the thin gate dielectric devices.
(b) Median measured, and simulated gate to channel leakage under inversion for PFETs with SiO_2 SDE spacer depending on the channel length at 85°C is shown.

Subthreshold Leakage

The subthreshold leakage is enhanced for the short channel PFETs with silicon oxide SDE spacer due to the higher DIBL effect (Fig. 5.37). The increase of the CET for the PFETs with silicon oxide spacer leads to increased DIBL (table 5.10), and subthreshold slope (table 5.12). The statistical data variations are due to the variance of the threshold voltage across the wafer (table 5.12).

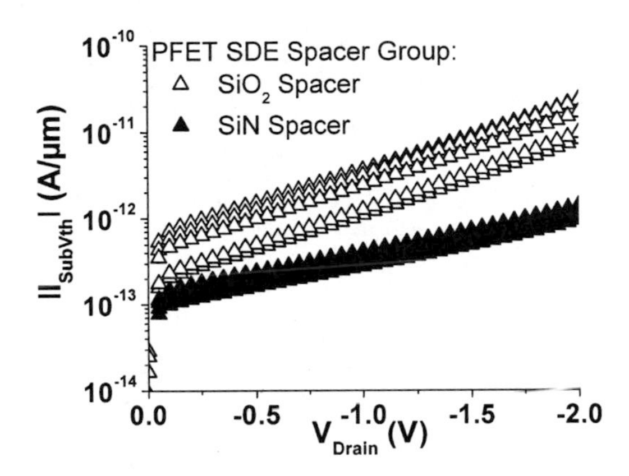

Figure 5.37: Subthreshold current at a gate bias of 0V of the 100nm long PFETs with thin gate dielectric at 85°C. Five dies are measured for each sample of the SDE spacer set.

5.5.4 Conclusion

This part of the chapter summarizes the results of different source/drain extension (SDE) spacer materials on the basic electrical characteristics, and leakage currents of the high-k PFET devices. Figure 5.29 illustrates that using a silicon oxide spacer results in the formation of undesirable bird beaks at the gate edge. That leads to an increase in average CET for the short channel devices. The standard silicon oxide spacers process can not be used for the production of high-k transistors with a silicon oxide interface layer of about 1nm. The effect is avoided when the source drain extension spacer consists of silicon nitride.

The higher permittivity of the silicon nitride spacer effects the electric field at the gate edge, and leads to changes in the gate induced drain leakage (GIDL). Silicon nitride is also known to have a higher silicon to spacer dielectric interface defect density [122, 123].

The PFETs with the silicon nitride SDE spacer show a reduction in overlap length (L_{ovl}) by about 1.5nm (table 5.12) which may be caused by small variation in spacer thickness

Table 5.13: Measured Leakage Current Ratios for Different SDE Spacer Materials

CET≈2nm	*Leakage Current Ratio at V_{Drain}=-1V and V_{Gate}=0V				
SDE Spacer	GIDL	SDE	Channel	Gate	Subthreshold
SiN/SiO_2 Spacer	3	1	1	106 (L_{Gate}=0.1μm) 1.5 (L_{Gate}=0.8μm)	1/3.3 (L_{Gate}=0.1μm) 1/1.2 (L_{Gate}=10μm)

CET≈5nm	*Leakage Current Ratio at V_{Drain}=-3V and V_{Gate}=0V				
SDE Spacer	GIDL	SDE	Channel	Gate	Subthreshold
SiN/SiO_2 Spacer	1.4	1	/	571 (L_{Gate}=0.8μm)	/

* Average measured leakage current ratio is given.

during processing. The variations in CET, and L_{ovl} also effects the basic electrical characteristics, such as threshold voltage or subthreshold slope (table 5.12), and the leakage currents.

The source/drain-, source/drain extension-, and channel current of the PFETs are not dependent on the different SDE spacer materials. The gate leakage, and gate overlap leakage for the PFETs with silicon nitride spacer devices is higher, and the subthreshold current is reduced. These changes in leakage are caused by the CET variations due to the bird beaks phenomenon of the transistors with silicon oxide SDE spacer.

The measured GIDL current increase for the PFET with silicon nitride SDE spacer is mainly caused by the lower interlayer thickness at the gate edge. The simulations additionally reveal, an increase in the slope of the GIDL due to the higher electric fields if a silicon nitride spacer is used. The influence of the changes in interface defect density with spacer dielectric could not be clarified in the simulations, because the statistical variations of the CET, and L_{ovl} have a larger effect on the final results. This leads to the conclusion that the change in concentration of the interface defects, responsible for generating the GIDL, must be below a factor of 2 for the two SDE spacer materials.

The results of the leakage current measurements of the thin gate dielectric devices are summarized in table 5.13. The main leakage current, measured at conditions equivalent to the operation at -1V at the drain and 0V at the gate, is the GIDL. Also subthreshold current, and the channel leakage contribute significantly to the off-current of short, and long channel devices, respectively.

For the thick gate dielectric PFETs, an unusual high gate leakage is found that points towards a process issue. The gate leakage of the samples with silicon nitride SDE spacer is reduced by optimizing the etch process (see also Fig. C.21 of the appendix). The GIDL increase with the change in SDE spacer material is less severe for the thick - compared to the thin gate dielectric PFETs. This finding is probably caused by the greater effect of bulk silicon traps on the overall GIDL. The results are summarized in table 5.13. Again the main leakage current is the GIDL, measured under operating equivalent conditions at -3V on the drain and 0V at the gate.

In order to merge the benefits of both materials, a PFET with a SDE spacer consisting of about 4nm silicon oxide, and 6nm silicon nitride is processed (see also table C.3 of the appendix). The first results show that the oxide regrowth is significantly reduced, leading to a gate current density independent of the channel length (Fig. 5.38). In this promising SDE spacer configuration, the GIDL current is expected to be reduced compared to the samples with pure silicon nitride SDE spacer. Further samples should be processed to investigate the SDE spacer set with about 4nm silicon oxide, and 6nm silicon nitride.

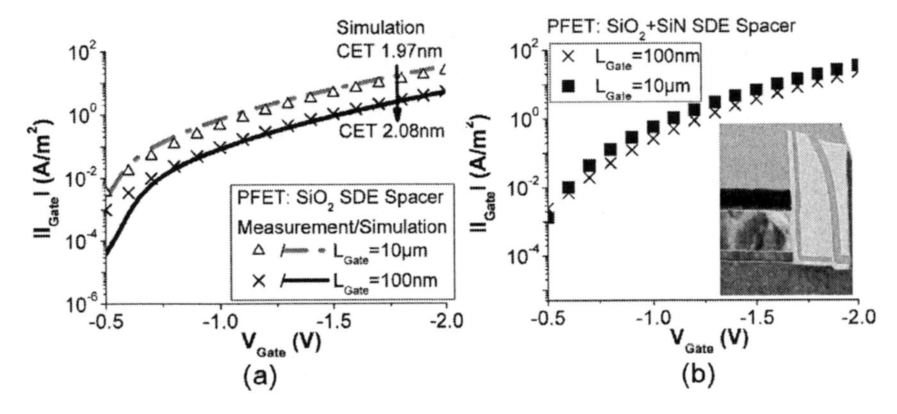

Figure 5.38: Measured median gate leakage under inversion of PFETs with thin gate dielectric depending on the channel length at 85°C. Measured curves are shown of samples with a:
(a) SiO_2 SDE spacer.
(b) Pile of SiO_2, and SiN SDE spacer (TEM analysis [112]).

5.6 Germanium Implantation during Gate Patterning

During gate patterning, the high-k gate stack is etched from above the source/drain area by using a combination of dry-, and wet processes. To remove the high permittivity material without residuals, it has to be in its amorphous state [114]. Therefore, a germanium ion implantation step with a dose of $3.0 \cdot 10^{14}$ atm/cm^2, and an energy of 10keV is used for high-k amorphization. Germanium is implanted directly at the unprotected gate edge [126]. The defect density, and the leakage currents are expected to be increased.

5.6.1 Description of Experiment

A smaller germanium implant energy of 3keV is tested, and compared to the standard 10keV implant. The basic electrical transistor properties are investigated to ensure the complete high-k removal.

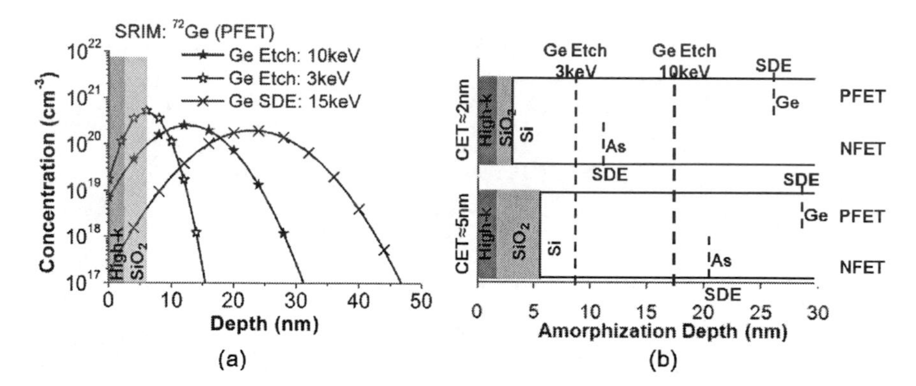

Figure 5.39: Implantation profiles of arsenic, and germanium calculated by SRIM [104]. (a) Analyzing thick gate dielectric PFETs, the germanium profiles from the etching step are compared to the profile due to SDE implantation.
(b) Estimated amorphization depth caused by the germanium implant prior to the high-k etch, and by the arsenic doping and germanium implantation during the SDE implantation.

SRIM simulations are done to calculate the germanium implantation profiles after the different process steps [104] (Fig. 5.39(a)). The implantation profiles, concentration (C_{Atom}) over depth (x), are calculated from the ion range (R_P), ion straggle (ΔR_P), and the implantation dose (N_{Dose}) (equation 5.1). The amorphization depth of the implants is approximated adding up the calculated ion range, and ion straggle [105]. In Figure 5.39, the high-k amorphization implant prior to the gate dielectric etch is compared to the source/drain extension implant. During SDE formation, the silicon substrate is amorphized by germanium ions for the PFETs, and by the arsenic implant for the NFETs.

$$C_{Atom}(x) = \frac{N_{Dose}}{\sqrt{2\pi} \cdot \Delta R_P} \cdot exp\left(\frac{(x - R_P)^2}{2 \cdot (\Delta R_P)^2}\right) \tag{5.1}$$

Interstitial defects gather at the amorphous/crystalline interface [95]. Figure 5.39 shows the advantage of the lower implantation energy. The interstitial defects are closer to the surface, where they can dissolve by out diffusion [98] during annealing. For the thick gate dielectric devices with a 3keV etching implant, the maximum of germanium ions lies within the gate dielectric which is removed in the following process step. From the SRIM results, the thick gate dielectric transistors are expected to show the maximum improvement in the trap assisted leakage current with the lower germanium implant energy for etching.

5.6.2 Electrical Characterization of Transistor Performance

The effect of the germanium implantation during the high-k etch on the basic electrical parameters is investigated. The most important parameters for the thin gate dielectric transistors are summarized in table 5.14. A comparison of the long channel thin-, and thick gate dielectric devices can be found in tables C.14, C.15 of the appendix.

Table 5.14: Basic Characteristics of MOSFETs with Different High-k Etches (CET≈2nm)

	L_{Gate}	PFET		NFET	
		10keV Ge	3keV Ge	10keV Ge	3keV Ge
CET (nm)	0.1μm	2.27±0.13	2.30±0.25	3.17±0.09	3.22±0.11
	10μm	2.15±0.02	2.20±0.03	1.98±0.01	2.00±0.01
L_{ov} (nm)	0.1μm	3.0±0.2	3.6±0.4	7.3±0.5	6.8±0.1
	10μm	4.0±0.55	4.0±0.5	4.6±0.7	3.8±1.5
C_{Diode} (fF/μm^2) V_{Diode}=-1V (CV*)	/	1.47±0.02	1.45±0.02	1.21±0.07	1.21±0.04
V_{th} (V) V_{Drain}=-0.05V (CV**)	0.1μm	-0.69±0.07	-0.68±0.09	0.40±0.05	0.40±0.05
	10μm	-0.57±0.08	-0.58±0.08	0.40±0.05	0.37±0.08
V_{th} (V) V_{Drain}=-0.05V (IV)	0.1μm	-0.66±0.01	-0.67±0.02	0.43±0.05	0.45±0.09
	10μm	-0.62±0.01	-0.63±0.01	0.46±0.02	0.48±0.10
V_{th} (V) V_{Drain}=-1V (IV)	0.1μm	-0.59±0.01	-0.58±0.01	0.41±0.01	0.42±0.02
V_{fb} (V) (CV**)	0.1μm	0.50±0.05	0.52±0.09	-0.82±0.08	-0.77±0.08
	10μm	0.45±0.05	0.45±0.05	0.74±0.06	0.74±0.06
I_{on} (μA/μm)	0.1μm	46.1±3.3	50.5±12.4	263±28	243±42
	10μm	0.64±0.03	0.64±0.07	4.4±0.4	4.4±0.01
S_{Vth} (mV/dec) V_{Drain}=-0.05V	0.1μm	90.8±0.4	95.3±43	102.8±6.1	98.3±2.4
	10μm	80.3±0.3	81.1±0.6	87.4±13.6	85.4±1.4
λ (IV***)	0.1μm	0.08±0.01	0.09±0.01	0.027±0.045	0.028±0.019

All values are given for measurements on 5 dies (wafer center). All devices have a SiN SDE spacer

* Capacitance voltage measurements are done on a diode structure with SDE and S/D implants.

** V_{fb} and V_{th} are taken from the point of maximum, and minimum slope of the CV.

*** DIBL parameter is determined after the V_{th} shift of $I_{Drain}V_{Gate}$ (section 2.2).

Changes in the basic electrical transistor parameters are within the statistical measurement variations for different germanium implant energies. So, it is concluded that the high-k is effectively amorphized, and removed with the 3keV implant. An increase in average CET, for the short channel- compared to the long channel NFETs, is found by the split capacitance voltage measurement. The observation is consistent with the decrease in gate leakage for these transistors which is reported in section 5.4.5.

5.6.3 Leakage Current Dependence on the Gate Patterning Process

The changes of the leakage currents due to the variation of the germanium etching implant will be analyzed in this section. Variations of the defect density in the different regions of the transistors are taken into account.

Source/Drain Leakage

The source/drain current is generated in the depletion region close to the metallurgical junction at about 70nm (PFET), and 80nm (NFET) from the silicon surface (Fig. 5.13(a)). A trap assisted tunneling via defects is modeled, using the theory developed by Hurkx et al. [38], causes the leakage of both NFETs, and PFETs.

Figure 5.40 shows that no difference of the effective electric field of the junctions is found within the measurement variations. But, the S/D leakage current is increased with higher germanium implant energy for etching. The increase is in average a factor of 5 for the PFETs, and the NFETs at ±3V. That leads to the conclusion, that more interstitial defects remain in the silicon substrate after the 10keV germanium amorphization implant, and enhance the leakage current.

Interstitial defects gather around the amorphous/crystalline interface, after the high-k amorphization- and source/drain implant [95]. Extended end of range defects should be healed out after the thermal treatment but a small amount of interstitials can diffuse further down in the silicon substrate [97]. These interstitials lead to an increased diode leakage.

Source/Drain Extension Leakage

The SDE current is generated inside the extension depletion region (Fig. 5.13(a)). It is a trap assisted tunneling current that can be described using the Hurkx model [38]. The PFET SDE leakage is mainly generated at the interface traps at the gate edge (N_{itedge}) (Fig. 5.16(b)). The current generated at defects around the former amorphous/crystalline interface (N_I) is about a factor 2 lower than the current generated at the N_{itedge}. The

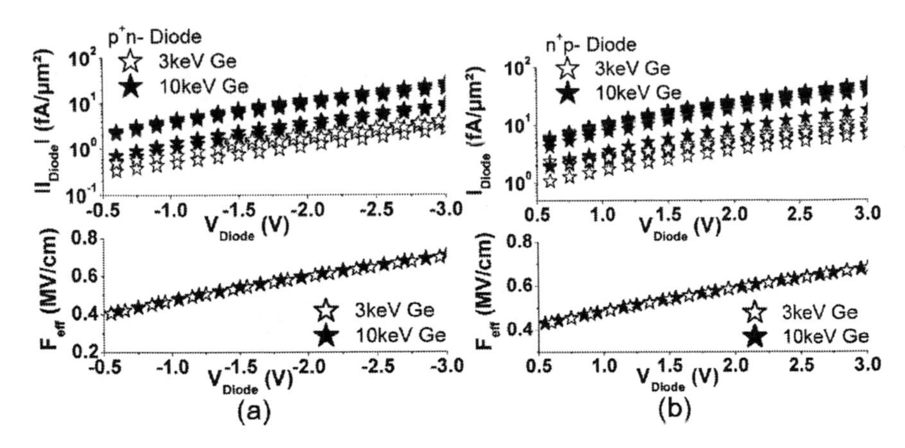

Figure 5.40: Measured S/D leakage current (lower part of the graph), and median electric field (upper part of the figure) at 85°C. The measurement of five diodes with a 10keV and and a 3keV germanium etching implant is shown in the upper part of the figure. The source/drain processing is analogues to the devices with thick gate dielectrics, analyzing:
(a) P^+n-junctions.
(b) N^+p diodes.

NFETs SDE current shows the same trend as the NFET source/drain current (Fig. 5.17), and is therefore not discussed in detail.

The SDE current is reduced for both thin-, and thick gate dielectric PFETs with lower germanium implant energy. Figure 5.41(a) shows the simulated, and average measured SDE current ratio which is calculated for the leakages of the thin gate dielectric PFETs. The ratio reduces from low to high voltages. From the simulation, a N_{itedge} defect ratio of 16.5 is found, rising germanium implant energy for high-k removal from 3keV to 10keV. The ratio of the bulk silicon traps N_I in the simulation is 4. A similar N_I ratio of 3.5 is found for the thick gate dielectric PFETs.

The amount of interstitial defects remaining in the bulk silicon (N_I) is expected to be smaller for the PFETs with 3keV germanium etch. Additionally, a huge reduction in silicon to dielectric interface traps at the gate edge (N_{itedge}) is found when a lower energy is used. The reduction in N_{itedge} is possibly explained by the lower lateral straggle of the germanium ions using 3keV amorphization implant. The channel width dependent SDE current is in the same order of magnitude as the the area dependent channel current from the N_{it} for the 10μm long PFETs. This leads to a germanium implant energy dependent current for the long channel PFETs when the bulk is biased (Fig. 5.42).

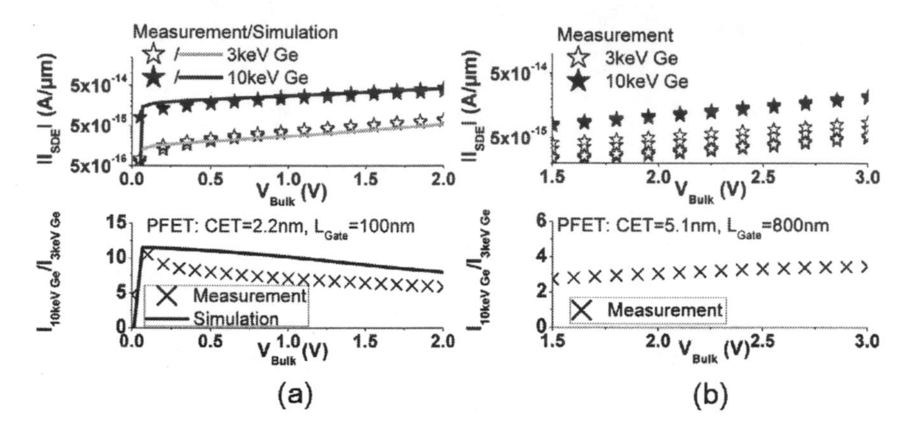

Figure 5.41: Measured, and simulated SDE leakage current of the five PFETs each at 85°C are shown in the upper part of the figure. The lower part of the graph presents the average leakage current ratio for the different germanium etching energies. Devices presented are transistors with a:
(a) Thin gate dielectric.
(b) Thick gate dielectric.

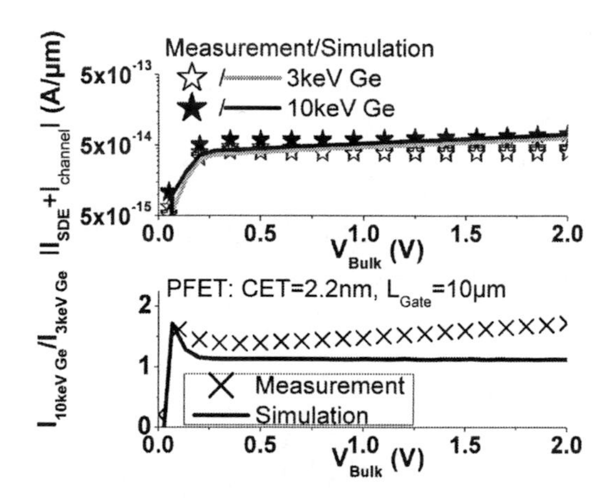

Figure 5.42: Measured, and simulated SDE- and channel leakage current of the five PFETs each at 85°C are shown in the upper part of the figure. The lower part of the graph presents the average leakage current ratio for the different germanium etching energies.

Gate Induced Drain Leakage

The gate induced drain leakage is a perimeter current that is generated in the gate to source/drain overlap region (Fig. 5.21(a)). The GIDL is caused by a trap assisted tunneling with phonon interaction (Hurkx) from the silicon to dielectric interface defects (N_{itedge}), and bulk silicon traps close to the surface ($N_{Isurface}$) [51]. The band to band tunneling leakage contribution to the total GIDL is increasing with the bias (section 5.4.4). A reduction in trap assisted GIDL due to the decrease in germanium implant energy for high-k preamorphization is found (Fig. 5.44, and 5.43).

The GIDL current of the NFET devices is relatively insensitive to the interface traps at the gate edge [50]. The interface trap assisted GIDL current is dominated by carrier generation at the $N_{Isurface}$ defects. The average GIDL ratio between the transistors with a high and low germanium energy is 3.5 for the thin gate dielectric NFETs, and fluctuates between 4 and 5 for the thick gate dielectric devices (Fig. 5.44(b), and 5.43(b)). As expected, the improvement for the thick gate dielectric NFETs is slightly better. Less germanium ions reach the silicon substrate during the implantation process if a thick gate dielectric is used. The result is similar to improvement in N_I found in the previous section.

For the thin gate dielectric PFETs, the trap assisted GIDL is dominated by N_{itedge}. For the thick gate dielectric PFETs, both N_{itedge} and $N_{Isurface}$ contribute to the trap assisted GIDL. The analysis of the SDE current has shown that the effect of the germanium preamorphization energy on the N_{itedge} is much higher compared to the $N_{Isurface}$. These finding are consistent with the high average GIDL current ratio with germanium preamorphization energy by a factor of 20 at -1V for the thin gate dielectric PFETs (Fig. 5.44(a)). The average leakage decrease of the thick gate dielectric PFETs is nearly constant at a factor of 1.3 (Fig 5.43(a)).

The simulations for the thin gate dielectric PFETs assume a trap band with constant defect density (N_{itedge}, Fig. 5.28), and a GIDL generated at a defect center with a constant trap energy. The simulated N_{itedge} ratio, that is found comparing 10keV and 3keV germanium implant samples, is 16. The average measured GIDL current ratio is very high at low voltages, and decreases exponentially at higher biases (Fig. 5.44(a)). The simulated GIDL current ratio also decreases with the bias, but the slope is lower. It is possible, that a two trap mechanism enhances the GIDL current for the PFETs with 10keV germanium implant energy. Another explanation could be a concentration gradient of N_{itedge} from the gate edge towards the middle of the channel.

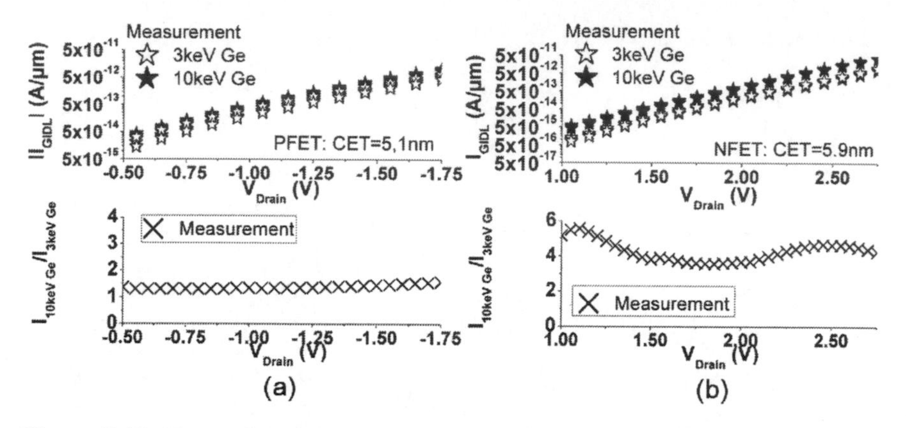

Figure 5.43: Measured, and simulated trap assisted GIDL current of five transistors each at 85°C are shown in the upper part of the figure. The lower part of the graph presents the average leakage current ratio for the different germanium implant energies. Thick gate dielectric devices presented are:
(a) PFETs.
(b) NFETs.

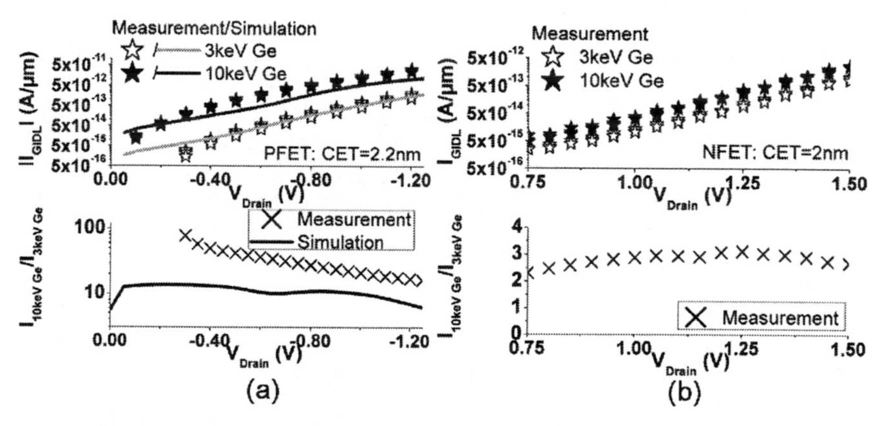

Figure 5.44: Measured, and simulated trap assisted GIDL current of five transistors each at 85°C are shown in the upper part of the figure. The lower part of the graph presents the average leakage current ratio for the different germanium implant energies. Thin gate dielectric devices presented are:
(a) PFETs.
(b) NFETs.

Gate Leakage

The germanium implant during the gate stack patterning does not effect the area dependent gate to bulk leakage, and gate to channel leakage (Fig. 2.2(a)). A band to band tunneling current causes the gate to source drain overlap current (section 5.4.5). The NFET gate overlap leakage is increased with higher germanium implant energy in average by a factor between 1.5 and 2.0. The increase is nearly independent of the gate voltage, and is not effected by the gate dielectric thickness (Fig. 5.45). In contrast, the PFET overlap leakage is unchanged or slightly reduced. However, the PFET gate leakage shows a higher die to die variation as the NFET.

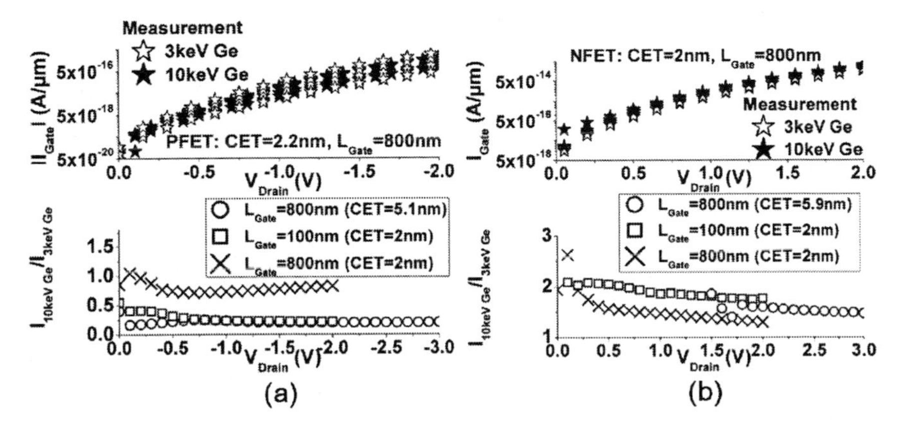

Figure 5.45: Measured gate overlap leakage of 800nm long transistors at 85°C is shown in the upper part of the figure. Five dies are measured for each sample. The lower part of the graph presents the average leakage current ratio for the different germanium implant energies. The thin gate dielectric devices presented are:
(a) PFETs.
(b) NFETs.

The gate overlap current is probably effected by charged traps in the gate dielectric, and at the silicon to gate oxide interface. Charged defects alter the electric field at the gate edge [127]. Negative charged defects in the gate dielectric would increase the NFET - and decrease the PFET overlap current. The increase in the NFET overlap leakage could also be explained by the interface trap enhanced tunneling mechanism, where interface traps act as a stepping stone for the tunneling process [128].

5.6.4 Conclusion

The effect, of the germanium implantation for high-k preamorphization during gate patterning, on the electrical transistor characteristics is summarized in this part of the chapter. The germanium implant energy is varied between 10keV, and 3keV. The maximum damage of the silicon substrate is closer to the silicon surface when a lower germanium energy is used (Fig 5.39).

No clear variations of the basic electrical transistor parameters due to the change from 10keV to 3keV germanium implant energy are found (table 5.14). The subthreshold leakage is not effected by the germanium implant. A high subthreshold current for the short channel NFETs due to the the low V_{th} of the devices, and the increase in DIBL with higher CET is measured.

In the gate patterning process, the germanium implant amorphizes the high-k, and the underlying silicon substrate. The silicon is amorphized again using another deeper germanium implant for the PFETs, and the arsenic doping implant for the NFETs during the source/drain extension-, and the source/drain formation. Subsequently, interstitial defects gather at the amorphous/crystalline interface forming end of range defects [95]. Thermal treatment dissolves the end of range defects leaving smaller interstitial defect clusters [101]. The interstitials are healed out due to out diffusion to the silicon surface during annealing [98]. With the 3keV germanium implant for high-k preamorphization, the maximum of the damage before thermal treatment is closer to the depletion region where the gate induced drain leakage, and the source/drain extension leakage are generated. But the damage is also closer to the silicon surface, and healed out more efficiently.

For the thick gate dielectric devices, the maximum of the 3keV germanium implant damage remains in the gate dielectric, which is etched from the source/drain region after the amorphization step (Fig 5.39). The changes of all leakage currents with germanium amorphization energy are summarized in table 5.15.

The source drain current is decreased by a factor of 5 for the thick dielectric transistors with lower germanium implant energy. As expected, less interstitial defects remain in the silicon substrate when a 3keV germanium implant is used. A small amount of the interstitial defects diffuses from the former amorphous/crystalline interface after around 30nm down to the source/drain depletion region (metallurgical junction PFET at 70nm, and NFET at 80nm) during the thermal treatment.

The NFET source drain extension leakage (SDE) is not analyzed separately, because the source/drain leakage dominates the perimeter current. The defect concentration in the SDE depletion region is not much increased compared to the n^+p source/drain region. The domination of the perimeter current by the source/drain leakage is probably caused by the shallow SDE arsenic amorphization implant, leaving only a low number of interstitial defects after recrystallization.

The PFET SDE leakage is reduced for transistors with lower germanium implant energy. Deduced from simulations, the amount of interstitial related defects remaining in the

Table 5.15: Measured Leakage Current Ratios for Different Ge etches

CET≈2nm		*Leakage Current Ratio at V_{Drain}=-1V and V_{Gate}=0V				
Transistor	Implant	GIDL	SDE	Channel	Gate	Subthreshold
PFET	Ge(10keV)/Ge(3keV)	21	7	1	1/2	1/1.2
NFET	Ge(10keV)/Ge(3keV)	3.5	/	0.8	1.8	1.4

CET≈5nm		*Leakage Current Ratio at V_{Drain}=-3V and V_{Gate}=0V				
Transistor	Implant	GIDL	SDE	Channel	Gate	Subthreshold
PFET	Ge(10keV)/Ge(3keV)	2.1	3.5	/	1/4.8	/
NFET	Ge(10keV)/Ge(3keV)	3.3	/	/	1.5	/

* Average measured leakage current ratio is given.

silicon bulk is reduced by a factor of about 4, moving from a 10keV to a 3keV germanium implant. The reduction is independent of the dielectric thickness of the device. The SDE leakage for the thin gate dielectric PFETs is mainly caused by silicon to dielectric interface defects at the gate edge. These defects are significantly reduced, by about a factor of 16 (TCAD simulations), when a 3keV germanium implant is used. The reduction in implant energy reduces the lateral straggle of the germanium ions.

The gate induced drain leakage (GIDL) current decreases with reduced germanium implant energy for high-k preamorphization. The trap assisted NFET GIDL is mainly caused by silicon bulk defects close to the silicon surface [50]. It is decreased by a factor of 3.5 for the thin gate dielectric devices, and a factor of 4.5 for the thick gate dielectric devices. A significant band to band tunneling contribution to the total GIDL is measured at 1.5V to 2V, and 2.6V to 3V for the thin, and thick gate dielectric NFETs, respectively.

Silicon to dielectric interface defects at the gate edge are the main generation center of the GIDL for the thin gate dielectric PFETs. The GIDL results are qualitatively reproduced in the simulations with the assumptions made for the SDE leakage calculation. But the GIDL current ratio, for the different germanium implant energies, is not constant in the trap assisted regime as expected by the simulations. A significant defect concentration gradient along the channel, or a two trap generation mechanism possibly alters the GIDL current. Only a small GIDL current reduction, of a factor of 1.3, with germanium implant energy is measured at the PFETs with thick gate dielectric. For the thick gate dielectric PFETs, charge carriers generated at both bulk defects close to the silicon surface, and silicon to dielectric interface defects contribute to the overall leakage.

The gate induced drain leakage is the main off-current for all PFETs of this sample set. For the thick gate dielectric PFETs with a 3keV germanium implant, the GIDL is 172pA/μm at -3V, which is close to the upper leakage current target 135pA/μm of the 65nm technology. Gate leakage, channel leakage, and subthreshold current are still major

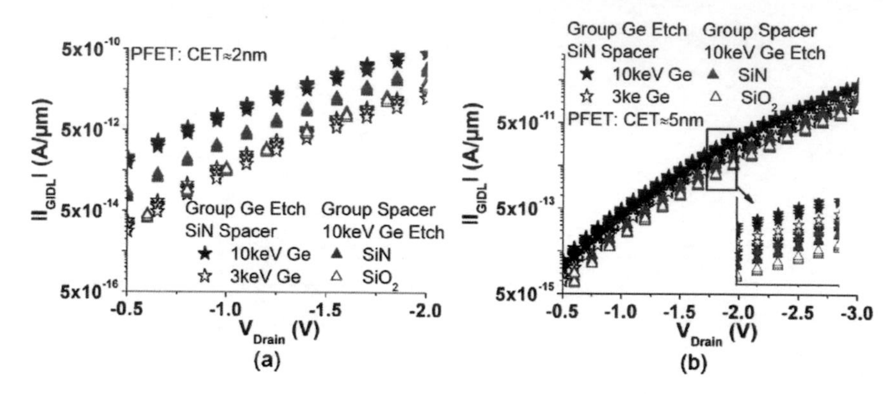

Figure 5.46: PFET GIDL current depending on the gate patterning, and - encapsulation of the two sample sets at 85°C. Measurements of five dies for each sample are presented analyzing:
(a) Thin gate dielectric PFETs.
(b) Thick gate dielectric PFETs.

issues for the thin gate dielectric NFETs at a gate bias of 0V, and a drain bias of +1V. The GIDL controls the off-current of the thick gate dielectric NFETs at a gate bias of 0V, and a drain bias of +3V. The GIDL of thick gate dielectric NFETs with a 3keV germanium implant is 33pA/μm at a bias of 3V, which meets the upper leakage current target 100pA/μm of the 65nm technology.

The sample set has a silicon nitride SDE spacer, to avoid unintended bird beaks. Comparing the sample set with different etches to the SDE spacer sample set, the etching process, and temperature budget is improved. This improvement leads to a reduced gate leakage, source/drain leakage, SDE leakage, and interface defect density in the channel middle. The GIDL is increased by this process changes. The use of smaller germanium implant energy for high-k removal reduces the GIDL. For the thin gate dielectric PFETs, the low GIDL level of the transistor with silicon oxide SDE spacers from the previous sample set is reached (Fig. 5.46).

6 Summary and Outlook

The continuous improvement in CMOS technology requires transistor scaling while maintaining acceptable leakage currents. Many process changes have been implemented to meet the scaling goals of the ITRS 2009 [24] for the DRAM peripheral MOSFETs (table 1.1).

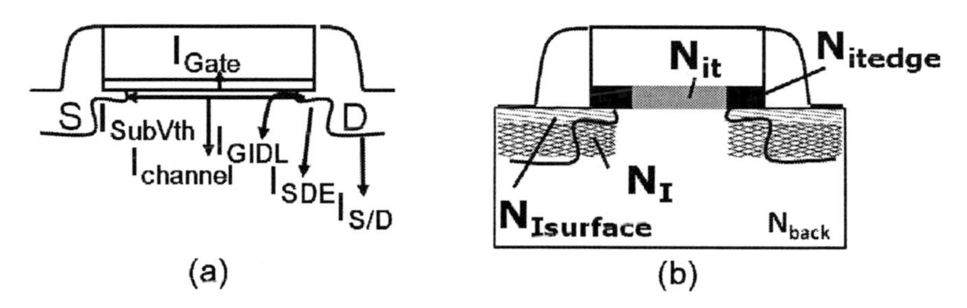

Figure 6.1: Schematic sketch of the intrinsic leakage currents, and traps responsible for generating MOSFET leakage.
(a) Leakage current occurrence in the MOSFET.
(b) Trap distribution of the transistors assumed in the simulation.

This study gives a complete overview of how particular process changes effect the intrinsic leakage currents of peripheral MOSFETs (Fig. 6.1(a)). Two types of transistors are investigated. Thick gate dielectric devices with a capacitance equivalent oxide thickness (CET) of about 5nm which are operated at ±3V, and thin gate dielectric devices with a CET of about 2nm which are operated at ±1V. The different leakage current paths of these MOSFETs are analyzed. The defect distribution in the different transistor regions is determined by electrical measurements, and simulations.

The effects of the following processes are investigated in separate sample sets, respectively: Carbon implantation in the source/drain junction for doping diffusion control, variation of the doping concentration in the channel and halo, and the high-k dielectric integration with special attention on the silicon nitride encapsulation and on the removal of the high-k during gate patterning.

6.1 Leakage Currents and Defect Distribution

Various factors effect the leakage currents which are listed in figure 6.1(a): The effective electric field, the charge carrier concentration, the trap energy, and the trap concentration. Temperature dependent leakage current measurements are analyzed, and simulations with TCAD Sentaurus [54] and ngspice [55, 74] are done, to determine the dominant leakage current mechanisms. For the defect assisted leakage currents, the best agreement between measurement and simulation is reached if the model of Hurkx et al. [38] is used. The model includes thermal generation of charge carriers at one main defect center, and electric field enhancement. Different defect regions are determined (Fig. 6.1(b)). The defect characteristics of various devices depend strongly on the process sequence and thermal budget, and can be found in the corresponding chapters. A summary of the dominant transistor leakage current mechanisms, and the associated trap centers is given in table 6.1.

Table 6.1: Summary of Intrinsic MOSFET Leakages

<table>
<tr><td rowspan="2">Leakage</td><td colspan="2">PFET</td><td colspan="2">NFET</td></tr>
<tr><td>CET≈5nm</td><td>CET≈2nm</td><td>CET≈5nm</td><td>CET≈2nm</td></tr>
<tr><td>$I_{S/D}$</td><td>Hurkx (N_{back})</td><td>-</td><td>Hurkx (N_{back})</td><td>-</td></tr>
<tr><td rowspan="2">I_{SDE}</td><td rowspan="2">Hurkx (N_I, $N_{Isurface}$)</td><td rowspan="2">Hurkx (N_I, N_{itedge} $N_{Isurface}$)</td><td rowspan="2">Hurkx (N_{back})</td><td>Hurkx (N_{back})</td></tr>
<tr><td>BTBT</td></tr>
<tr><td>$I^*_{channel}$</td><td colspan="2">Hurkx (N_{it})</td><td colspan="2">Hurkx (N_{it}, N_{back})</td></tr>
<tr><td rowspan="2">I_{GIDL}</td><td>Hurkx (N_{itedge}, $N_{Isurface}$)</td><td>Hurkx (N_{itedge})</td><td colspan="2">Hurkx ($N_{Isurface}$)</td></tr>
<tr><td colspan="4">BTBT</td></tr>
<tr><td>I_{Gate}</td><td colspan="4">FN- and direct tunneling</td></tr>
<tr><td rowspan="2">I_{SubVth}</td><td colspan="4">Diffusion current</td></tr>
<tr><td colspan="4">DIBL</td></tr>
</table>

* Gate at depletion bias
Hurkx - Tunneling enhanced generation leakage [38]
BTBT - Band to band tunneling in silicon [38]
Fowler Nordheim tunneling (FN), direct tunneling - Band to band tunneling through the gate dielectric [26]
DIBL - Drain induced barrier lowering [36]

6.2 Carbon in the Junction Extension

Carbon is implanted in the PFET source/drain junction to reduce the transient enhanced boron doping diffusion. The implant energy is 4keV, and a dose of $3.5 \cdot 10^{14}$atm/cm^2

or $4.0{\cdot}10^{14}$atm/cm^2 is used. A standard silicon oxynitride device is investigated. The chosen variation in carbon implant dose is small. No change in the electric field of the source/drain junction is found by capacitance voltage measurements. The determined variation of the overlap length is below 1nm. It is concluded that a dose of $3.5{\cdot}10^{14}$atm/cm^2 is sufficient to effectively suppress transient enhanced diffusion. The effect of the carbon dose on the leakage currents is summarized in table 6.2.

An increase in the defect density at the gate edge is found. The bulk silicon defects (N_I) in the extension depletion region are increased by a factor of 1.35 for a carbon dose increase of 1.15. One conclusion of this result could be that every extra carbon leads to roughly two electrical active defects. It is possible that carbon stabilizes small interstitial defect cluster during the thermal treatment [108]. Another explanation for the increased defect density is the formation of carbon interstitial traps, or phosphorus substitutional pairs [109]. The influence of carbon induced defects does not reach the source/drain depletion region.

Additionally, a higher carbon dose increases the interface defect density at the gate edge (N_{itedge}) by at least a factor of 1.3. The increase is higher for the thick gate dielectric PFETs. Different methods are used for the gate oxide formation of both devices: A thermal oxidation process for the thick gate dielectric PFETs, and an in situ steam generated oxide for the thin gate dielectric PFETs. The dielectric interface quality in the channel middle is found to be better for the thick gate dielectric PFETs (Fig. 4.30). Different impacts of the gate stack etching procedure on thick, and thin gate dielectric transistor interfaces possibly cause the diverse effect of carbon on the N_{itedge} of both devices.

Table 6.2: Influence of Junction Implants on the Leakage Currents of the PFETs

Implantation Dose	Device	$I_{S/D}$	I_{SDE}	$I_{channel}$	I_{GIDL}	I_{Gate}	I_{SubVth}
C↑	thick/thin	↔	↑	↔	↑	↔	↔
V_{th} ↓	thick	↔	↓	↓	↓	↔	↑
V_{th} ↓	thin	↔	↓	↔	↔	↔	↑
Halo ↓	thick	↔	↓	↔	↓	↔	↑ (short channel) ↔ (long channel)
Halo ↓	thin	↔	↓	↔	↔	↔	↑ (short channel) ↔ (long channel)

↔ No effect of implantation dose change on leakage current is measured.
Results in the table are given for implant dose variations around 20%.

6.3 V_{th}- and Halo Implant

The V_{th} implant leads to a higher doping concentration beneath the gate, and increases the threshold voltage [3]. The halo implant is used for short channel control. It decreases the depletion width of the S/D junction by a higher bulk doping directly at the gate edge [8]. For short channel devices, the halo doping effects the threshold voltage. Both implants are reduced for the high-k gate stack to reach the threshold voltage targets. The reduction effects the electric field, and consequently the leakage currents. Small variations in V_{th}- and halo implantation dose (below 20%) are investigated. The effect on the PFET leakage currents is summarized in table 6.2. Transistors with silicon oxynitride as a gate dielectric are used for this investigation.

The bulk doping in the source/drain depletion region is mainly defined by the well implants. Tails of the halo implant change the effective field at the S/D junction (Fig. 4.40). But the alteration of source/drain leakage due to defect density variation is higher than the changes with the electric field. Small variations of the V_{th}- and halo implants do not effect source/drain leakage.

As expected, the source/drain extension leakage is reduced with lower halo implant. The gate induced drain leakage (GIDL) of the thick gate dielectric transistors is also reduced. The GIDL of the thin gate dielectric devices is unaffected by small changes in implants, if the overlap length remains nearly constant. The trap assisted GIDL of the thin gate dielectric PFETs is caused by the band bending at the deep depletion, and is generated at interface traps at the gate edge (N_{itedge}). The band bending at the interface in the overlap region depends mainly on the source/drain doping rather than on the bulk doping.

The subthreshold current is increased when V_{th}- and halo implants are reduced, but the DIBL effect reacts sensitively to the oxide thickness [37]. In scaled devices, the better channel control with reduced CET will partially compensate the disadvantage of the reduced V_{th}- and halo implantation doses.

6.4 Silicon Oxynitride vs. Hafnium Silicon Oxide Gate Dielectric

High permittivity dielectrics are used to reach a low CET of the gate stack. The improvement in gate leakage due to an increased physical thickness of the gate dielectric is shown in figure 6.2. The MOSFETs with hafnium silicon oxide are compared to 65nm standard silicon oxynitride transistors. Not only the gate dielectric material but also the doping implants are varied between the samples.

The high-k NFETs reach the ITRS target for the threshold voltage of 0.4V using a polysilicon gate electrode [24, 131]. An increase in CET with reducing channel length is observed for the high-k NFETs. The edge effect is possibly related to the high-k polysilicon interlayer. For the PFETs, a titanium nitride top electrode is used to reduce the gate electrode depletion [12]. The effective work function of the titanium nitride leads to a high absolute threshold voltage of the PFETs [113]. Therefore, an aluminum oxide layer is deposited on top of the high-k. Charged aluminum defects are driven into the high-k by a high

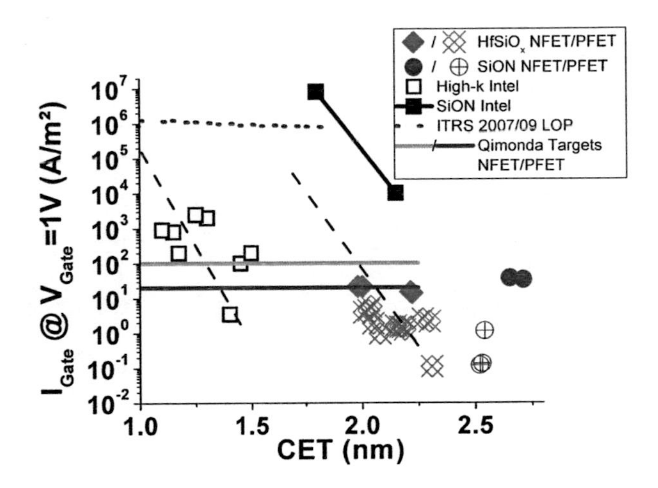

Figure 6.2: Measured gate leakage versus CET at 85°C in comparison with Intel results, and the ITRS targets for low operational power (LOP) devices [24, 129, 130]. Qimonda targets are defined for the peripheral transistors of the 46nm technology.

temperature anneal [113]. Threshold voltages between -0.6V to -0.7V are extracted for the PFETs.

The use of aluminum defects to tune the threshold voltage leads to an increased interface defect density at the conduction band edge of the high-k PFETs compared to the silicon oxynitride reference devices (table 5.4). But if the silicon oxynitride PFETs are scaled down to a similar CET, a rise in interface defect density at the conduction band edge is also found (Fig. 4.6). The defect increase for the scaled silicon oxynitride PFETs can be explained by boron penetration from the polysilicon in the gate dielectric, and the subsequent formation of charged defects.

Fig 6.3 summarizes the leakage currents, comparing 65nm high-k technology with the silicon oxynitride reference. The gate leakage of the thin high-k devices is reduced. Further improvement of the isolation processes of the thick gate dielectric transistors is needed to reduce the gate leakage of the high-k devices to the expected values. The subthreshold leakage increases with lower V_{th}, and higher CET. The short channel high-k NFET will probably meet the Qimonda off-current criteria of 2nA/μm when the CET is decreased. The interface trap assisted GIDL is increased for the high-k PFETs due to the change in source/drain extension spacer material, the decrease in CET, and the necessity of an additional germanium implant during gate patterning. The variations of the leakage currents with the extension spacer material, and the germanium implantation are investigated in more detail.

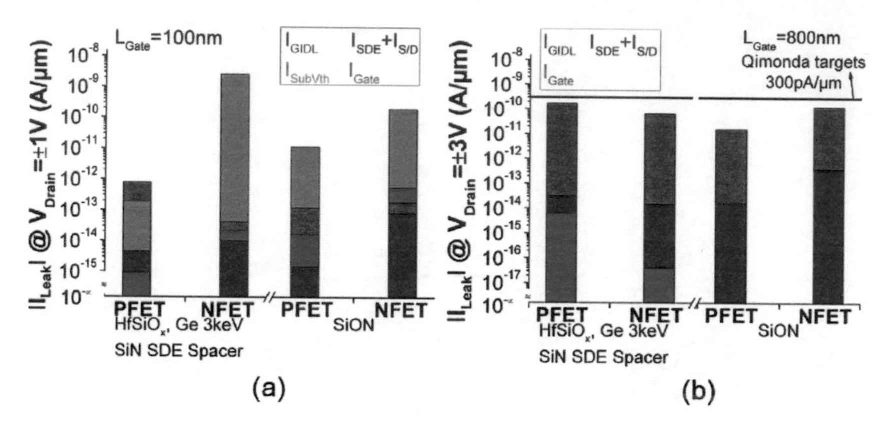

Figure 6.3: Composition of total leakage current of high-k, and silicon oxynitride devices at 85°C, and a gate bias of 0V. Qimonda targets are given under standby conditions assuming the GIDL to be the main leakage in the future generations. Results of:
(a) Transistors with a CET of 2nm to 3nm are shown.
(b) MOSFETs with a CET of 5nm to 6nm are presented.

6.4.1 Silicon Oxide vs. Silicon Nitride Extension Spacer

Silicon oxide spacers are used in Qimonda standard technology to define the source/drain to the gate overlap, and to protect the gate edge during ion implantation steps. High-k processing leads to undesirable bird beaks at the silicon oxide interlayer towards the gate edge (Fig 5.29). The bird beaks phenomenon is characterized by an increase in CET, and decrease in gate leakage density with lower channel length. Silicon nitride extension spacers are used to avoid this effect. The effect of the extension spacer material on the leakage currents is summarized in table 6.3.

Simulations reveal that with the use of silicon nitride as extension spacer material, the lateral electric field increases at the dielectric to silicon interface. This leads to an increase in trap assisted tunneling GIDL. For the PFETs, a higher GIDL current for the devices with silicon nitride extension spacer is measured. The source/drain extension current in these samples is generated mainly at the bulk silicon traps (N_I), and is therefore not effected. In comparing measurement and simulation, no clear indication is found for the expected increase of electrical active defects at the spacer to silicon interface.

The decrease in oxide thickness at the gate edge for the PFETs with silicon nitride extension spacer increases the GIDL-, gate current, and decreases the subthreshold leakage.

6.4.2 Germanium Implantation for Gate Patterning

To remove the gate stack from the source/drain area, the hafnium silicon oxide needs to be preamorphized [114]. A germanium implant is used for this purpose. Two different implant energies of 3keV and 10keV are tested. Both energies are sufficient to remove the high-k without residuals. The lower energy is less destructive, and places the damage closer to the silicon surface. The 3keV implant damage is expected to be healed out more efficiently during the thermal treatment [98]. The effect of the germanium implantation energy on the leakage currents is summarized in table 6.3.

As expected, the interstitial trap density in the extension depletion region is reduced with lower germanium implant energy. Also, fewer interstitials diffuse to the source/drain depletion region, improving the leakage. The maximum of the damage is placed closer to the silicon surface, nevertheless the defect concentration at the dielectric interface (N_{itedge}), and at bulk close to the surface ($N_{Isurface}$) is significantly reduced. N_{itedge} is decreased by a factor of 16 for the thin gate dielectric devices, probably due to the lower lateral straggle of the 3keV germanium implant.

The NFET gate overlap leakage is decreased. For the PFET, a small increase or unchanged gate leakage within the statistical variations is found.

Table 6.3: Influence of High-k Integration on the Leakage Currents of the MOSFETs

Integration	Device	$I_{S/D}$	I_{SDE}	$I_{channel}$	I_{GIDL}	I_{Gate}	I_{SubVth}
SiN/SiO_2 SDE Spacer	thin/thick	↔	↔ (N_I) ↑(N_{itedge})	↔	↑	↑	↓
Ge Dose ↓ Amorphization	thin/thick	↓	↓	↔	↓	Area:↔ Overlap: ↓ (NFET) ↑ (PFET)	↔

↔ No effect of process change on leakage current is measured.

6.5 Outlook

To reduce the GIDL current without forming unintended bird beaks, an extension spacer consisting of a silicon oxide and silicon nitride layer is tested. First results are promising. The influence of the thickness of the two layers on the leakage should be investigated in more detail.

The absolute value of the threshold voltage of the metal gate PFET is too high. This too high threshold voltage is caused by the work function of the titanium nitride electrode. The effect is partly compensated by an aluminum oxide tuning layer in the gate stack. Still the PFETs do not reach the ITRS targets (table 1.1).

The NFET polysilicon high-k gate stack shows edge effects, leading to an increased CET with reducing channel length. The effect is possibly related to the interlayer between polysilicon, and hafnium silicon oxide. Further process improvements are necessary, to delay the introduction of an expensive metal gate for the high-k NFET of peripheral DRAM devices.

7 Bibliography

[1] J. van der Spiegel. *Introduction to nanoscale science and technology III: From microscale to nanoscale devices.* Springer US, 2004.

[2] G.E. Moore. Cramming more components onto integrated circuits. *Electronics Magazine*, 39(8):114, 1965.

[3] Y. Taur and T.H. Ning. *Fundamentals of modern VLSI devices.* Cambridge University Press, 1998.

[4] R.H. Dennard, F. H. Gaensslen, H.-N. Yu, V.L. Rideout, E. Bassous, and A.R. Leblanc. Design of ion-implanted mosfets with very small physical dimensions. *IEEE Journal of solid state circuits*, 9(5):256–268, 1974.

[5] P. Chatterjee, W.R. Hunter, T.C. Holloway, and Lin Y.T. The impact of scaling laws on the choice of n-channel or p-channel for mos vlsi. *IEEE Electron Device Letters*, 1:220–223, 1980.

[6] G. Baccarani, M.R. Wordeman, and R.H. Dennard. Generlized scaling theory and its application to a 1/4 micrometer mosfet design. *IEEE Transaction on Electron Devices*, 31:452, 1984.

[7] S. Ogura, P.J. Tsang, W.W. Walker, D.L. Critchlow, and J.F. Shepard. Design and characteristics and the lightly doped drain-source insulated gate field-effect transistor. *IEEE Journal of solid state circuits*, 15(4):424–432, 1980.

[8] S. Ogura, C.F. Codella, N. Rovedo, J.F. Shepard, and J. Riseman. A half micron mosfet using double implanted ldd. In *International Electron Device Meeting*, pages 718–721, 1982.

[9] G.D. Wilk, R.M. Wallace, and J.M. Anthony. High κ gate dielectrics: Current status and materials properties considerations. *Journal of Applied Physics*, 89(10):5243–5273, 2001.

[10] W.-J. Qi, R. Nieh, B. H. Lee, L. Kang, Y. Jeon, K. Katsunori, T. Ngai, S. Banerjee, and J.C. Lee. Moscap and mosfet characteristics using zro_2 gate dielectric deposited directly on si. In *International Electron Device Meeting*, pages 145–148, 1999.

[11] K. Mistry et al. A 45nm logic technology with high-k+metal gate transistors, strained silicon, 9 cu interconnect layers, 193nm dry patterning, and 100% pb-free packaging. In *International Electron Device Meeting*, pages 247–250, 2007.

[12] J. Robertson. High dielectric constant gate oxides for metal oxide si transistors. *Reports on Progress in Physics*, 69:327–396, 2006.

[13] NEC Electronics. Ultimatelowpower technology, Reading: 2011. http://www2.renesas.com/process/en/lowpower_ overview.html.

[14] K. Roy, S. Mukhopadhyay, and H. Mahmoodi-Meimand. Leakage current mechanisms and leakage reduction techniques in deep submicrometer cmos circuits. *Proceedings of the IEEE*, 91(2):305–327, 2003.

[15] S. Mukhopadhyay and K. Roy. Modeling and estimation of total leakage current in nano-scaled cmos devices considering the effect of parameter variation. In *Proceedings of the International Symposium on Low Power Electronics and Design*, pages 172–175, 2003.

[16] A. Keshavarzi, K. Roy, and C.F. Hawkins. Intrinsic leakage in deep submicron cmos ics- measurement-based test solutions. *IEEE Transactions on Very Large Scale Intergration (VLSI) Systems*, 8(6):717–723, 2000.

[17] T. Schloesser et al. A $6f^2$ buried wordline dram cell for 40nm and beyond. In *International Electron Device Meeting*, pages 1–4, 2008.

[18] T. Vogelsang. Understanding the energy consumption of dynamic random access memories. In *International Symposium on Microarchitecture*, pages 363–374, 2010.

[19] Y.G. Shin, K.-J. Nam, H. Hwang, J.H. Han, S. Hyun, S. Choi, and J.-T. Moon. Gate oxide effect on wafer level reliability of next generation dram transistors. In *IEEE International Reliability Physics Symposium*, pages 282–286, 2010.

[20] D. Park, W. Lee, and B.-I. Ryu. Stack dram technology for the future. In *International Symposium on VLSI Technology, Systems, and Applications*, pages 1–4, 2006.

[21] Smithsonian: The Chip Collection. Dram technology, Reading: 2011. http://smithsonianchips.si.edu/ice/cd/MEMORY97/SEC07.PDF.

[22] Elpida. Elpida uses high-k metal gate technology to develop 2-gigabit ddr2 mobile ram. Press Release, Reading: 2011. http://www.elpida.com/en/news/2011/06-15.html.

[23] J.W. Faul and D. Henke. Transistor challenges - a dram perspective. *Nuclear Instruments and Methods in Physics Research B*, 237:228–234, 2005.

[24] Semiconductor Industry Association. *International Technology Roadmap of Semiconductors (ITRS 2009)*, 2009.

[25] M. Luisier and A. Schenk. Two-dimensional tunneling effects on the leakage current of mosfets with single dielectric and high-κ gate stacks. *IEEE Transactions on Electron Devices*, 55(6):1494–1501, 2008.

[26] J.C. Ranuarez, M.J. Deen, and C.-H. Chen. A review of gate tunneling current in mos devices. *Microelectronics Reliability*, 46:1939–1956, 2006.

[27] S.M. Sze. *Physics of semiconductor devices (2nd edition)*. John Wiley & Sons, 1981.

[28] D.K. Schroder. *Semiconductor material and device characterization (3rd Edition)*. John Wiley & Sons, 2006.

[29] M.J.J. Theunissen and F.J. List. Analysis of the soft reverse characteristics of n+p source drain diodes. *Solid States Electronics*, 28(5):417–424, 1985.

[30] S.D. Khanin. Hopping electronic conduction in metal oxide films and their insulating properties. In *IEEE Proceedings of the Conference on Conduction and Breakdown in Solid Dielectrics*, pages 57–61, 1992.

[31] O. Engstroem, B. Raeissi, J. Piscator, I.Z. Mitrovic, S. Hall, H.D.B. Gottlob, M. Schmidt, P.K. Hurley, and K. Cherkaoui. Charging phenomena at the interface between high-k dielectrics and sio_x interlayers. *Journal of Telecommunications and Information Technology*, 1:81–90, 2010.

[32] A. Weber. *Charakterisierung von Leckstrompfaden in DRAM Speicherzellen und deren Reduktion*. Phd, Fakultät Elektrotechnik der Universität Hamburg-Harburg, 2007.

[33] R. Entner. *Modeling and simulation of negative bias temperature instability*. Phd, Technische Universität Wien, 2007.

[34] C. Svensson and I. Lundstroem. Trap-assisted charge injection in nmos structures. *Journal of Applied Physics*, 44:4657–4663, 1973.

[35] K.M. Cao, W.-C. Lee, W. Liu, X. Jin, P. Su, S.K.H. Fung, J.X. An, B. Yu, and C. Hu. Bsim4 gate leakage model including source-drain partition. In *IEEE International Electron Device Meeting Technical Digest.*, pages 815–818, 2000.

[36] R.R. Troutman. Vlsi limitation from drain-induced barrier lowering. *IEEE Journal of Solid-State Circuits*, 14(2):383–391, 1979.

[37] S.G. Chamberlain and S.R. Ramanan. Drain-induced barrier-lowering analysis in vsli mosfet devices using two dimensional numerical simulations. *IEEE Transaction on Electron Devices*, ED-33(11):1745–1753, 1986.

[38] Klaasen D.B.M. Hurkx, G.A.M. and and M.P.G. Knuvers. A new recombination model for device simulation including tunneling. *IEEE Transactions on Electron Devices*, 39(2):331–338, 1992.

[39] B. van Zeghbroeck. *Principles of semiconductor devices*. University of Colorado, 2007.

[40] B. Somanathan Nair and S.R. Deepa. *Solid State Devices*. PHI Learning Private Limited, 2010.

[41] R.N. Hall. Electron-hole recombination in germanium. *Physical Review*, 87:387, 1952.

[42] W. Shockley and W.T. Read. Statics of recombination of holes and electrons. *Physical Review*, 87:335, 1952.

[43] de Graaff H.C. Hurkx, G.A.M. and, W.J. Kloosterman, and M.P.G. Knuvers. A new analytical diode model including tunneling and avalanche breakdown. *IEEE Transactions on Electron Devices*, 39(9):2090–2098, 1992.

[44] E.O. Kane. Theory of tunneling. *Journal of Applied Physics*, 32(1):83–91, 1961.

[45] A.G. Chynoweth, W.L. Feldmann, and R.A. Logan. Excess tunnel current in silicon esaki junctions. *Physical Review*, 121(3):684–694, 1961.

[46] A. Schenk. Rigorous theory and simplified model of band-to-band tunneling in silicon. *Solid State Electronics*, 36(1):19–34, 1993.

[47] A. Czerwinski, E. Simoen, C. Claeys, K. Klimaf, D. Tomasszewski, J. Gibki, and J. Katckia. Optimized diode analysis of electrical silicon substrate properties. *Journal of the Electrochemical Society*, 145:2107–2112, 1998.

[48] G. Roll, M. Goldbach, and L. Frey. Leakage current and defect characterization of p^+n-source/drain diodes. *Microelectronic Reliability*, 51(12):2081–2085, 2011.

[49] A. Bouhdada, S. Bakkali, and A. Touhami. Modelling of gate-induced drain leakage in relation to technological parameters and temperature. *Microelectronic Reliability*, 37(4):649–652, 1997.

[50] I.-C. Chen, C.W. Teng, D.J. Coleman, and A. Nishimura. Interface-trap enhanced gate-induced leakage current in mosfet. *IEEE Electron Device Letters*, 10(5):216–218, 1989.

[51] F. Gilibert, D. Rideau, A. Dray, F. Agut, M. Minondo, A. Juge, P. Masson, and R. Bouchakour. Characterization and modeling of gate-induced-drain leakage. *IEICE Transactions on Electron Devices*, E88(5):829–837, 2005.

[52] T. Hori. Drain-structure design for reduced band-to-band and band-to-defect tunneling leakage. In *IEEE Technical Digest of the Symposium on VLSI Technology*, pages 69–70, 1990.

[53] S. Döring (Namlab gGmbH). Personal Communication.

[54] Synopsys. *Sentaurus device user guide (version D-2010.03)*, 2010.

[55] University of California. *NGSPICE user manual (ngspice-rework-17 version 0.2)*, 1996.

[56] H. Haddara. *Characterization Methods for submicron MOSFETs*. Kluwers, Academic Publishers, 1995.

[57] M.A. Pavanello, J.A. Martino, E. Simoen, and C. Claeys. Impact of halo implantation on 0.13 μm floating body partially depleted soi n-mosfets in low temperature operation. *Solid-State Electronics*, 49:1274–1281, 2005.

[58] A. Poyai, E. Simoen, C. Claeys, R. Rooyackers, G. Badenes, and E. Gaubas. High purity silicon vi: Lifetime and leakage currrent studies in shallow p-n junctions fabricated in a deep high-energy boron implanted well. *Proceedings of the Electrochemical Society*, 17:403–413, 2000.

[59] A.S. Grove and D.J. Fitzgerald. Surface effects on p-n junctions: Characteristics of surface space-charge regions under non-equilibrium conditions. *Solid-State Electronics*, 9:783–806, 1966.

[60] J. Huang, T.P. Chen, M.S. Tse, and C.H. Ang. Characterization of interface degradation in deep submicron mosfets by gated- controlled- diode measurements. *Microelectronics Journal*, 33:639–643, 2002.

[61] K. Romanjek, F. Andrieu, T. Ernst, and G. Ghibaudo. Characterization of the effective mobility by split c(v) technique in sub 0.1 μm si and sige pmosfets. *Solid-State Electronics*, 49:721–726, 2005.

[62] R. van Langevelde and M. Klaassen. Effect of gate-field dependent mobility degradation on disortion analysis in mosfets. *IEEE Transactions on Electron Devices*, 44(11):2044–2052, 1997.

[63] S.-I. Takagi, A. Toriumi, M. Iwase, and H. Tango. On the universality of inversion layer mobility in si mosfet's: Part i-effects of substrate impurity concentration. *IEEE Transactions on Electron Devices*, 41(12):2357–2362, 1994.

[64] J.S. Brugler and P.G.A. Jespers. Charge pumping in mos devices. *IEEE Transactions on Electron Devices*, ED-16(3):297–302, 1969.

[65] J.L. Autran, B. Balland, and G. Barbottin. *Instabilities in silicon devices 3: New insulator, devices and radiation effects*. Elsvier Book, 1999.

[66] G. Groseneken, H.E. Maes, N. Beltran, and R.F. de Keersmaecker. A reliable approach to charge-pumping measurements in mos transistors. *IEEE Transactions on Electron Devices*, ED-31(1):42–53, 1984.

[67] P. Masson, J.-L. Autran, and J. Brini. On the tunneling component of charge pumping current in ultrathin gate oxide mosfets. *IEEE Electron Device Letters*, 20(2):92–94, 1999.

[68] M. Gaitan, E.W. Enlow, and T.J. Russell. Accuracy of the charge pumping technique for small geometry mosfets. *IEEE Transactions on Nuclear Science*, 36(6):1990–1997, 1989.

[69] C.-C. Lu, K.-S. Chang-Liao, C.-Y. Lu, S.-C. Chang, and T.-K. Wang. Leakage effect suppression in charge pumping measurement and stress-induced traps in high-k gated mosfets. In *IEEE International Semiconductor Device Research Symposium*, 2007.

[70] M.B. Zahid, R. Degraeve, J.F. Zhang, and G. Groseneken. Impact of process conditions on interface and high-k trap density studied by variable t_{charge}-$t_{discharge}$ charge pumping (vt^{2}cp). *Microelectronic Engineering*, 84:1951–1955, 2007.

[71] S. Jakschik. *Charakterisierung alternativer Dielektrika fuer zukuenftige dynamische Halbleiterspeicher.* Phd, Fakultät Elektrotechnik und Informationstechnik der TU Dresden, 2004.

[72] W. Chen, A. Balasinski, and T.P. Ma. Lateral profiling of oxide charge and interface traps near the mosfet junctions. *IEEE Transactions on Electron Devices*, 40(1):187–196, 1993.

[73] Synopsys. *Sentraurus process user guide (version C-2009.06)*, 2009.

[74] T.H. Morshed, W. Yang, M.V. Dunga, X. Xi, J. He, W. Liu, Kanyu, M. Cao, X. Jin, J.J. Ou, M. Chan, A.M. Niknejad, and C. Hu. *BSIM4.6.4 MOSFET model.* University of California, 2009.

[75] K. Mothes (Qimonda Dresden). Personal Communication.

[76] S. W. Lim, T.-Y. Luo, and J. Jiang. Mechanism of silicon dioxide decoupled plasma nitridation. *Japanese Journal of Applied Physics*, 45(15):L413–L415, 2006.

[77] Y. Song, H. Zhou, and Q. Xu. Source /drain technologies for scaling of nanoscale cmos device. *Solid State Sciences*, 13:294–305, 2011.

[78] N. Cagnat, C. Laviron, D. Mathiot, B. Duriez, J. Singer, R. Gwoziecki, F. Salvetti, B. Dumont, and A. Pouydebasque. Ultra shallow junction optimization with non doping species co-implantation. *Proceedings of the Material Research Symposium*, 912:0912–C01–02, 2006.

[79] P. Pichler and D. Stiebel. Current status of models for transient phenomena in dopant diffusion and activation. *Nuclear Instruments and Methods in Physics Research B*, 186:256–274, 2002.

[80] V. Moroz, Y.-S. Oh, D. Pramanik, D. Graoui, and M. Foad. Optimizing boron junctions through point trap and stress engineering using carbon and germanium co-implants. *Applied Physical Letters*, 87:051908, 2005.

[81] C.F. Tan, E.F. Chor, H. Lee, J. Liu, E. Quek, and L. Chan. Leakage suppression of gated diodes fabricated under low-temperature annealing with substitutional carbon si$-1-y$c$_y$ incorporation. *IEEE Electron Device Letters*, 26(4):252–254, 2005.

[82] B.J. Pawlak, T. Janssens, B. Brijs, W. Vandervorst, E.J.H. Collart, S.B. Felch, and N.E.B. Cowern. Effect of amorphization and carbon co-doping on activation and diffusion of boron in silicon. *Applied Physical Letters*, 89:062110, 2006.

[83] S. Mirabella, A. Coati, D. de Salvador, E. Napolitani, A. Mattoni, G. Bisognin, M. Berti, A. Carnera, A.V. Drigo, S. Scalese, S. Pulvirenti, A. Terrasi, and F. Priolo. Interaction between self-interstitials and substitutional c in silicon: interstitial trapping and c clustering mechanism. *Physical Review B*, 65:045209, 2002.

[84] D. Song, J. Lim, K. Lee, Y.-J. Park, and H.-S. Min. Optimization study of halo doped mosfets. *Solid-State Electronics*, 39(6):923–927, 1996.

[85] H. Hortenbach (Fresenius). Personal Communication.

[86] J.R. Pfiester, L.C. Parrillo, and F.K. Baker. A physical model for boron penetration through thin gate oxides from p^+ polysilicon gates. *IEEE Electron Device Letters*, 11(6):247–249, 1990.

[87] R.B. Fair. Modeling boron diffusion in ultrathin nitrided oxide p^+ si gate technology. *IEEE Electron Device Letters*, 18(6):244–247, 1997.

[88] D.L. Flowers. Defect generation in gate oxides during the polysilicon doping and activation process. *Microelectronic Engineering*, 14(1):1–12, 1991.

[89] Y. Son, S. Yand, B.C. Kim, J. Kim, C.-R. Moon, D. Lee, J.-D. Lee, B.-G. Park, and H. Shin. Extration of interface-states energy distribution in nitrided and pure gate dielectrics for metal oxide semiconductor field effect transistor applications. *Japanese Journal of Applied Physics*, 48:054502, 2009.

[90] W. Jellett, C. Zhang, H. Jin, P.J. Smith, and K.J. Weber. Boron emitters: Defects at the silicon-silicon dioxide interface. In *IEEE Photovolatic Specialists Conference 2008*, pages 1–6, 2008.

[91] M. Otani, K. Shiraishi, and A. Oshiyama. First-principle calculations of boron-related defects in sio_2. *Physical Review B*, 68:184112, 2003.

[92] Z. Wang, C.G. Parker, D.W. Hodge, R.T. Croswell, N. Yang, V. Misra, and J.R. Hauser. Effect of polysilicon gate type on the flatband voltage shift for ultrathin oxide-nitride gate stacks. *IEEE Electron Device Letters*, 21(4):170–172, 2000.

[93] Y.H. Lin, C.S. Lai, C.L. Lee, T.F. Lei, and T.S. Chao. Nitidization of stacked poly-si gate to supress the boron penetration in pmos. *IEEE Transactions on Electron Devices*, 43(7):1161–1165, 1996.

[94] I.V. Antonova and S.S. Shaimeev. Transformation of radiation defect clusters in b^+ ion-implanted silicon. *Physica Status Solidi (a)*, 153:329–336, 1996.

[95] P. Pichler and D. Stieble. Current status of models for transient phenomena in dopant diffusion and activation. *Nuclear Instruments and Methods in Physics Research B*, 186:256–274, 2002.

[96] P.H. Keys. *Phosphorus-defect interaction during thermal annealing of ion implanted silicon.* Phd, Universtiy of Florida, 2001.

[97] D. Girginoudi and C. Tsiarapas. Comparative study on eor and deep level defects in preamorphised si implanted with b^+, bf_2^+, and f^+-b^+. *Nuclear Instruments and Methods in Physics Research B*, 266:3565–3576, 2008.

[98] P.J. Bedrossian, N.J. Caturla, and T. Diaz de la Rubia. Damage evolution and surface defect segregation in low-energy ion-implanted silicon. *Applied Physics Letters*, 70(2):176–178, 1997.

[99] M. Pepper. An introduction to silicon inversion layers. *Contemporary Physics*, 18(5):423–454, 1977.

[100] A.O. Adan and K. Higashi. Off-state leakage current mechanism in bulk si and soi mosfets and their impact on cmos ulsis standby current. *IEEE Transaction on Electron Devices*, 48(9):2050–2057, 2001.

[101] J.L. Benton, K. Halliburton, S. Libertino, D.J. Eaglesham, and S. Coffa. Electrical signatures and thermal stability of interstitial clusters in ion implanted si. *Journal of Applied Physics*, 84(9):4749–4756, 1998.

[102] Y. Suprun-Belevich, L. Palmetshofer, B.J. Sealy, and N. Emerson. Mechanical strain and electrically active defects in si implanted with ge^+ ions. *Semiconductor Science Technology*, 14:565–569, 1999.

[103] I. Ban, M.C. Oetztuerk, K. Christensen, and D.M. Maher. Effects of carbon implantation on generation lifetime in silicon. *Applied Physics Letters*, 68(4):499–501, 1996.

[104] J.F. Ziegler, J.P. Biersack, and M.D. Ziegler. *SRIM- The stopping and range of ions in matter.* Pergamon Press, 1985.

[105] R.B. Fair. Damage removal/dopant diffusion tradeoffs in ultra-shallow implanted p^+-n junctions. *IEEE Transactions on Electron Devices*, 37(10):2237–2242, 1990.

[106] R.B. Beck, T. Brozek, J. Ruzyllo, S.D. Hossain, and R.E. Tressler. Effect of carbon and thermal oxidation of silicon and electrical properties of sio_2-si structures. *Journal of Electronic Materials*, 22(6):689–694, 1993.

[107] G. Roll, S. Jakschik, M. Goldbach, T. Mikolajick, and L. Frey. Carbon junction implant: Effect on leakage current and defect distribution. In *European Solid State Device Research Conference*, pages 329–332, 2010.

[108] P.S. Chen, T.E. Hsieh, and C.-H. Chu. Removal of end-of-range defects in ge^+-preamorphized si by carbon ion implantation. *Journal of Applied Physics*, 85(6):3114–3119, 1999.

[109] P. Pichler. *Intrinsic point defects, impurities, and their diffusion in silicon.* Computational Microelectronics, 2004.

[110] G. Roll, S. Jakschik, A. Burenkov, M. Goldbach, T. Mikolajick, and L. Frey. Impact of carbon junction implant on leakage currents and defects distribution: Measurement and simulation. *Solid States Electronics*, 65-66:170–176, 2011.

[111] R.B. Beck, T. Brozek, J. Ruzyllo, S.D. Hossain, and R.E. Tressler. Effect of carbon on thermal oxidation of silicon and electrical properties of sio_2-si structures. *Journal of Electronic Materials*, 22(6):689–694, 1993.

[112] A. Leuteritz (Qimonda Dresden). Personal Communication.

[113] W.W. Wang, K. Akiyama, W. Mizubayashi, T. Nabatame, H. Ota, and A. Toriumi. Effect of al-diffusion-induced positive flatband voltage shift on the electrical characteristics of al-incorporated high-k metal-oxide-semiconductor field-effective transistor. *Journal of Applied Physics*, 105:064108, 2009.

[114] D. Shamiryan, V. Paraschiv, M. Claes, and W. Boullart. Low substrate damage high-k removal after gate patterning. In *Defects in High-k Dielectrics: Proceedings of the NATO Advanced Research Workshop on Defects in Advanced High-k Dielectric*, pages 331–338, 2006.

[115] M. Mildner (Fraunhofer Dresden). Personal Communication.

[116] J.T. Ryan, P.M. Lenahan, G. Bersuker, and P. Lysaght. Electron spin resonance observation of oxygen deficient silicon atoms in the interfacial layer of hafnium oxide based metal-oxide-silicon structures. *Applied Physics Letters*, 90:173513, 2007.

[117] J.P. Han, E.M. Vogel, E.P. Gusev, E.C. Demic, C.A. Richter, D.W. Heh, and J.S. Suehle. Asymmetric energy distribution of interface traps in n- and p-mosfet with hfo_2 gate dielectric on ultrathin sion buffer layer. *IEEE Electron Device Letters*, 25(3):126–128, 2004.

[118] C.-T. Sah, J.Y.-C. Sun, and J.J.T. Tzou. Study of atomic models of three donorlike defects in silicon metal-oxide-semiconductor structures from their gate material and process dependencies. *Journal of Applied Physics*, 55(6):1525–1545, 1984.

[119] O. Ghobar, D. Bauza, and B. Guillaumot. Defects in interfacial layer sio_2-hfo_2 gate stacks: Depth distribution and identification. In *Proceedings of the International Integrated Reliability Workshop*, pages 94–98, 2007.

[120] D. Ranka, A.K. Rana, R.K. Yadav, and D. Giri. Performance analysis of fd-soi mosfet with different gate spacer dielectric. *International Journal of Computer Applications*, 18(5):22–27, 2011.

[121] S. Chang, H. Shin, and J. Lee. Off-state leakage currents of mosfets with high-κ dielectrics. *Journal of the Korean Physical Society*, 41(6):932–936, 2002.

[122] S. Mahapatra, V.R. Rao, B. Cheng, M. Khare, C.D. Parikh, J.C.S. Woo, and J.M. Vazi. Performance and hot-carrier reliability of 100nm channel length jet vapor deposited si_3n_4 mnsefets. *IEEE Transaction on electron devices*, 48(4):679–684, 2001.

[123] R. Degraeve, M. Cho, B. Govoreanu, B. Kaczer, M.B. Zahid, J. Van Houdt, M. Jurczak, and G. Groseneken. Trap spectroscopy by charge injection and sensing (tscis): a quantitative electrical technique for studying defects in dielectric stacks. In *IEEE International Electron Device Meeting Technical Digest*, page 775, 2001.

[124] H.R. Hurff, A. Hou, C. Lim, Y. Kim, J. Barnett, G. Bersuker, G.A. Brown, C.D. Yound, P.M. Zeitzoff, J. Gutt, P. Lysaght, M.I Gardner, and R.W. Murto. High-k gate stacks for planar, scaled cmos integrated circuits. *Microelectronic Engineering*, 69:152–167, 2003.

[125] S. Jansen (Fraunhofer Dresden). Personal Communication.

[126] G. Roll, S. Jakschik, M. Goldbach, A. Wachowiak, T. Mikolajick, and L. Frey. Analysis of the effect of germanium preamorphization on interface defects and leakage current for high-k metal-oxide-semiconductor field-effect transistor. *Journal of Vacuum Science and Technology B*, 29(1):01AA05, 2011.

[127] S.-G. Hu, Y. Hao, X.-H. Ma, Y.-R. Cao, C. Chen, and X.-F. Wu. Hot-carrier stress effects on gidl and silc in 90nm ldd-mosfet with ultra-thin gate oxide. *Chinese Physical Letters*, 26(1):017304, 2009.

[128] W.Y. Loh, B.J. Cho, and M.F. Li. Correlation between interface traps and gate oxide leakage current in the direct tunneling regime. *Applied Physics Letters*, 81(2):379–381, 2002.

[129] R. Chau, S. Datta, M. Doczy, J. Kavalieros, and M. Metz. Gate dielectric scaling for high-performance cmos: from sio2/polysi to high-k/metal-gate. In *International Workshop on Gate Insulators (IWGI)*, pages 124–126, 2003.

[130] Semiconductor Industry Association. *International Technology Roadmap of Semiconductors (ITRS 2007)*, 2007.

[131] F. Tadashi, N. Yasushi, T. Toshifumi, and I. Kiyotaka. A new high-k transistor technology implemented in accordance with the 55nm design rule process. *Technical Journal NEC*, 1(5):42–46, 2006.

[132] K. Muraoka, K. Kurihara, N. Yasuda, and H. Satake. Optimum structure of deposited ultra thin silicon oxynitride film to minimize leakage currents. In *Proceedings of the Electrochemical Society: Silicon nitride and silicon dioxide thin insulating films IV*, pages 418–433, 2003.

8 Personal Bibliography

Paper:

[48] G. Roll, M. Goldbach, and L. Frey. Leakage current and defect characterization p+n-source/drain diodes. *Microelectronics Reliability*, 51(12): 2081-2085, 2011.

[110] G. Roll, S. Jakschik, A. Burenkov, M. Goldbach, L. Frey and T. Mikolajick. Impact of carbon junction implant on leakage current and defect distribution: Measurement and Simulation. *Solid-State Electronics*, 65-66: 170-176, 2011.

[126] G. Roll, S. Jakschik, A. Wachowiak, M. Goldbach, L. Frey, and T. Mikolajick. Analysis of the effect of germanium preamorphization on interface defects and leakage current for high-k metal-oxide-semiconductor field-effect transistor. *Journal of the Vacuum Science and Technology B*, 29:01AA05, 2011.

Conferences:

G. Roll, M. Goldbach, S. Jakschik, A. Wachowiak, L. Frey. Gate edge optimization for LSTP high-k metal gate technology. *European Solid States Device Conference*, 2009 (poster).

G. Roll, S. Jakschik, A. Wachowiak, M. Goldbach, L. Frey and T. Mikolajick. Analysis of the effect of germanium preamorphization on interface defects and leakage current for high-k metal-oxide-semiconductor field-effect transistor. *Workshop on Dielectrics in Microelectronics*, 2010 (talk).

G. Roll, S. Jakschik, M. Goldbach, L. Frey and T. Mikolajick. Carbon junction implant: Effect on leakage currents and defect distribution. *Proceedings of the European Solid States Device Conference*, pages 329-332, 2010 (talk).

G. Roll, S. Jakschik, A. Wachowiak, M. Goldbach, L. Frey and T. Mikolajick. Intrinsic MOSFET Leakage of High-k Peripheral DRAM Devices: Measurement and Simulation. *International Symposium on VLSI Technology, Systems and Applications*, 2012 (talk).

List Of Abbreviations

Notation	Description
a/c	amorphous/crystalline
AC	Alternate Current
ALD	Atomic Layer Deposition
BTBT	Band to Band Tunneling
CET	Capacitance Equivalent Oxide Thickness
CMOS	Complementary Metal Oxide Semiconductor (Technology)
CP	Charge Pumping
CV	Capacitance Voltage
DC	Direct Current
DIBL	Drain Induced Barrier Lowering
ECB	Electron Conduction Band
EELS	Energy Electron Loss Spectroscopy
EOR	End of Range (Defects)
EOT	Equivalent Oxide Thickness
EVB	Electron Valence Band
FP	Frenkel Poole
GIDL	Gate Induce Drain Leakage
HVB	Hole Valence Band
ITRS	International Technology Roadmap for Semiconductors
IV	Current Voltage
LOP	Low Operational Power (Devices)
MOSFET	Metal Oxide Semiconductor Field Effect Transistor

Notation	Description
NFET	N-Channel Field Effect Transistor
PFET	P-Channel Field Effect Transistor
PNA	Post Nitridation Anneal
PVD	Physical Vapor Deposition
RTP	Rapid Thermal Processing
S/D	Source/Drain
SDE	Source/Drain Extension
SIMS	Secondary Ion Mass Spectroscopy
TEM	Transmission Electron Microscopy
TEOS	Tetra-Ethyl-Ortho-Silicate

List Of Symbols

Notation	Description
ΔR_P	Ion Straggle during Implantation
$\Phi_{ECB,SiO2}$	Barrier Height for Electron Conduction Band Tunneling through SiO_2
ϵ_0	Vacuum Permittivity: $8.85 \cdot 10^{-12}$F/m
ϵ_{SiO2}	Dielectric Constant of SiO_2: 3.9
ϵ_{Si}	Dielectric Constant of Silicon: 11.7
ϵ_{ox}	Dielectric Constant of the Gate Dielectric
η	Drain Induced Barrier Lowering Geometry Parameter
γ	Constant
$\hbar$	Planck Constant Scaled with Circular Constant: $h/(2\pi)$
λ	Drain Induced Barrier Lowering Parameter
μ_{eff}	Effective Mobility
μ_{max}	Maximum Mobility
ϕ_b	Barrier Height
π	Circle Constant:3.142
σ	Capture Cross Section
σ_e	Capture Cross Section Electron
σ_h	Capture Cross Section Holes
τ_{Diff}	Diffusion Lifetime
τ_{Gen}	Generation Lifetime
A	Area
a	Number of Measurements
b	Prefactor
B_{FN}	Slope of Gate Leakage by Fowler Nordheim plot
B_{FP}	Prefactor Frenkel Poole Current
B_{Hurkx}	Electric Field Factor for Diode Leakage by Hurkx Mechanism
C	Capacitance
C_{Atom}	Concentration of Atoms
C_{Diode}	Junction Capacitance

Notation	**Description**
C_{ov}	Parasitic Overlap Capacitance
C_{ox}	Gate Dielectric Capacitance
D_{CV}	Dissipation Factor of the CV Measurement
D_{Diff}	Diffusion Coefficient
D_{itedge}	Region of Increased Interface Trap Density at the gate edge of the MOSFET
D_{it}	Interface Trap Density per Energy and Area
E	Energy
e	Elementary Charge: $1.602 \cdot 10^{-19}$J
E_a	Activation Energy
E_{em}	Energy of Electron Emission
E_g	Energy of the Semiconductor Band Gap
E_{hm}	Energy of Hole Emission
E_i	Mid gap Energy of the Semiconductor Band Gap
E_{tVB}	Trap energy from the Valence Band
E_t	Trap Energy
f	Frequency
F_{eff}	Effective Electric Field
F_{ox}	Electric Field Over the Gate Dielectric
F_{Si}	Effective Electric.Field in Silicon
g_{sat}	Saturation Transconductance (V_{Gate}=±1V)
h	Planck constant: $6.26 \cdot 10^{-34}$Js
I	Current
I_{BTBT}	Band to Band Tunneling Current
I_{Bulk}	Bulk/Well Current
$I_{channel}$	Leakage Current from the Channel to the Bulk
I_{CP}	Charge Pumping Current
I_{Drain}	Drain Current
I_{Gate}	Leakage through the gate dielectric
I_{Gen}	Generation Current
I_{GIDL}	Gate Induced Drain Leakage
I_{Hurkx}	Leakage Current due to trap assisted tunneling with phonon interaction
I_{Leak}	Leakage Current
I_{off}	Transistor Off-Current (V_{Gate}=0V, V_{Drain}=±1V)
I_{on}	Transistor On-Current (V_{Gate}=V_{Drain}=±1V)
$I_{S/D}$	Source/Drain Leakage

Notation	**Description**
I_{Sat}	Saturation Current of the Junction
I_{SDE}	Source/Drain Extension Leakage
I_{SubVth}	Subthreshold Leakage Current
I_{Total}	Total Current
J_A	Area Dependent Current Density
J_P	Current divided by Perimeter Length
k	Boltzmann Constant: $1.381 \cdot 10^{-23}$J/K
L_{Gate}	Gate Length
L_{ov}	Overlap Length
m	Transistor Body factor: $(S_{Vth} \cdot e)/(2.3kT)$
m^*	Effective Mass
$m_{electron}$	Electron Mass: $9.109 \cdot 10^{-31}$kg
N	Si Trap Density
N_{back}	Background Trap Density of the Transistor Bulk
N_{Dop}	Doping Concentration of Silicon
N_{Dose}	Ion Implantation Dose
$N_{Isurface}$	Region of Increased Trap Density around the S/D surface in the MOSFET
N_{it}	Interface Trap Density per Area
N_I	Region of Increased Trap Density around the S/D in the MOSFET
n_i	Intrinsic Carrier Concentration of Silicon
P	Perimeter Length
Q_{acc}	Accumulation Charge
Q_i	Inversion Charge
r	Fitting Parameter of the Diode CV Curves Related to Abrupt (0.5) or Linear Graded Junction (0.33)
R_P	Ion Range during Implantation
S_{Vth}	Subthreshold Slope
T	Temperature
t_{ox}	Physical Gate Dielectric Thickness
t_{fall}	Pulse Fall Time
t_{rise}	Pulse Rise Time

Notation	Description
t_{SiO2}	Physical thickness Silicon Oxide
V	Voltage
V_{Amp}	Pulse Amplitude
V_{Bulk}	Bulk/Well Voltage
V_{Diode}	Diode Voltage
V_{Drain}	Drain Voltage
V_{fb}	Flatband Voltage
V_{Gate}	Gate Voltage
V_{int}	Fitting Parameter of the Diode CV Curves Related to the Build In Potential
$V_{S/D}$	Voltage Difference from Source/Drain
V_{Supply}	Supply Voltage of DRAM Memory
V_{th0}	MOSFET Threshold Voltage at low Drain Bias (V_{Drain}=±0.05V)
V_{th}	Threshold Voltage
v_{th}	Thermal Velocity
W	Width of the Depletion Region
W_{Gate}	Width of the MOSFET Gate
x	Position
Y	Y-Function for V_{th} Determination
y	Position
z	Measurement Quantity

Index

A Appendix: Measurement Setup

A.1 Test Structures

Table A.1: Geometry of the Devices with Different Junction Implants

Test Structure	Type	CET (nm)	L_{Gate} (μm)	W_{Gate}(μm)	Amount
MultiFET	PFET	≈2	0.058	5	17339
	PFET	≈2	0.065	5	15325
	PFET	≈2	0.1	5	9911
	PFET/NFET	≈2	5	5	200
Overlap	PFET/NFET	≈2	0.8	148.8	42
MultiFET	PFET	≈5	0.124	5	8055
	PFET/NFET	≈5	5	5	200
Overlap	PFET	≈5	0.8	48	260
	NFET	≈5	0.8	48	200
Test Structure	Type	Device	A_{Diode} (μm^2)	P_{Diode}(μm)	Amount
Diode 1	p^+n/n^+p	CET≈5nm	5610.8	3871.5	1
Diode 2	p^+n/n^+p	CET≈5nm	7230.2	37735.0	1

Table A.2: Geometry of the Devices with High-k Metal Gate

Test Structure	Type	CET (nm)	L_{Gate} (μm)	W_{Gate}(μm)	Amount
MultiFET	PFET/NFET	≈2	0.1	5	8280
	PFET/NFET	≈2	10	5	108
	PFET/NFET	≈2	1	5	1000
Overlap	PFET/NFET	≈2	0.8	148.8	42
$Overlap^{Op}$	PFET/NFET	≈2	0.8	62.5	200
Overlap	PFET/NFET	≈5	0.8	62.5	200
Test Structure	Type	Device	A_{Diode} (μm^2)	P_{Diode}(μm)	Amount
Diode 1	p^+n/n^+p	CET≈5nm	5605.5	3758.7	1
Diode 2	p^+n/n^+p	CET≈5nm	7272.4	38356.2	1

^{Op} Transistors with nitride and oxide spacer which were produced with immersion lithography.

A.2 Temperature Control of the Measurements

The temperature calibration of the measurement setup is investigated using a thermal-couple (table A.3, Fig. A.1).

Table A.3: Temperature Calibration with Thermocouple on Chuck

Set T (°C)	Measured T (°C)
150	131
85	75
55	49
40	36
25	24
10	11

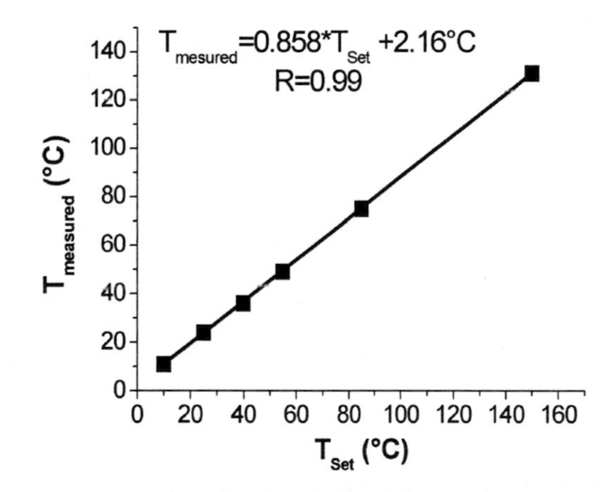

Figure A.1: Linear correlation of measured and set temperature of the chuck.

A.3 Accuracy of the Measurement Setup at High Frequencies

Measurement accuracy of the setup is tested at room temperature, and 500kHz. The current of 100fA, and a capacitance of 20fF can be measured accurately (Fig A.2). Reliable results can be obtained down to $1 \cdot 10^{-16}$A/μm, and $2 \cdot 10^{-17}$F/μm for a PFET with a gate length of 5μm of the MultiFET structure. Frequency dependent capacitance voltage measurement show a reliable measurement window between 40kHz and 2MHz (Fig. A.3). Reliable bias pulses are obtained at the device under test down a pulse period of 500ns, and with a rise time of 100ns (Fig. A.4).

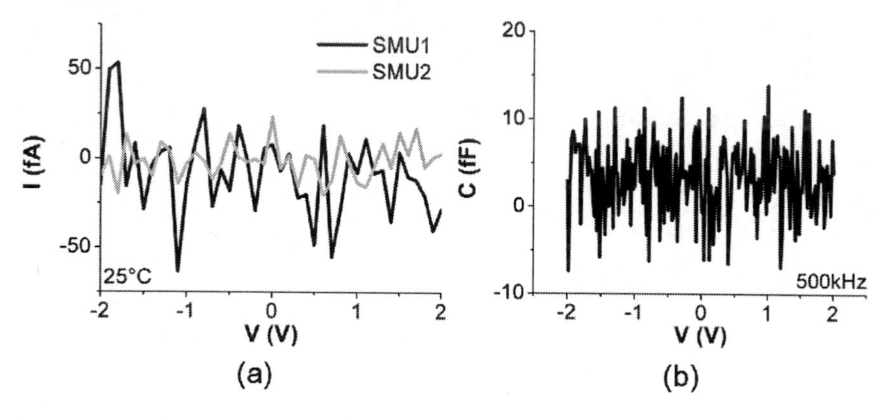

Figure A.2: Noise of the measurement setup at 25°C including cables and needles.
(a) Noise measured at different source measurement units (SMUs).
(b) Noise determined at the capacitance voltage unit for 500kHz, and 50mV AC amplitude.

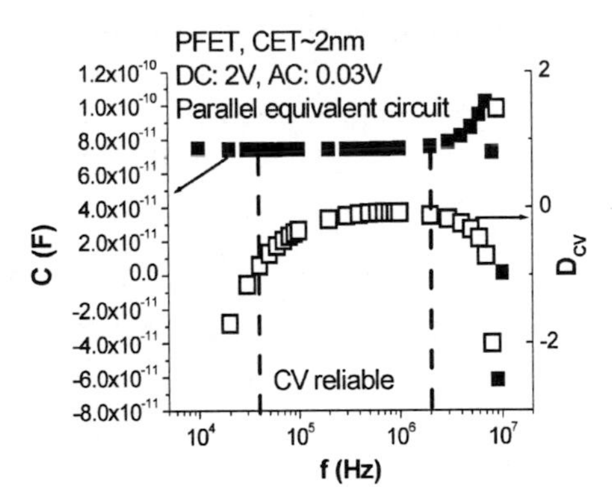

Figure A.3: Frequency dependent capacitance voltage measurements for a PFET in accumulation. Reliable CV data can be obtained between 40kHz and 2MHz.

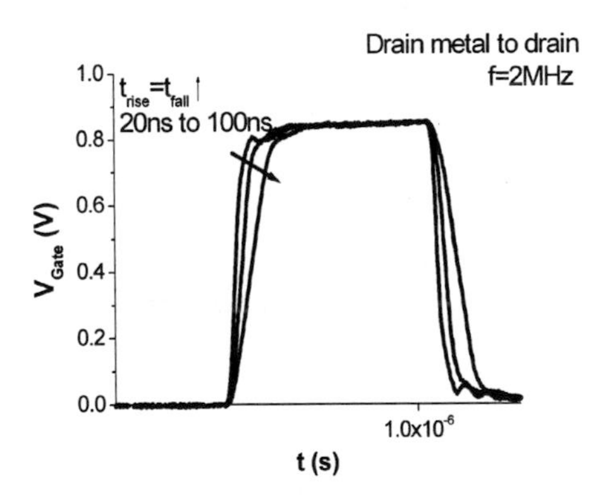

Figure A.4: Pulses measured when pulse generator and oscilloscope are shorted over the drain metalization. Reliable pulses are measured for rise- and fall times of 100ns at a frequency of 2MHz.

A.4 Charge Pumping Measurements

Comparison of CP base sweeps at different test structures (Fig. A.5). A MultiFET with a gate width of 5μm, and an Overlap transistor with a gate width of 148.8μm are measured.

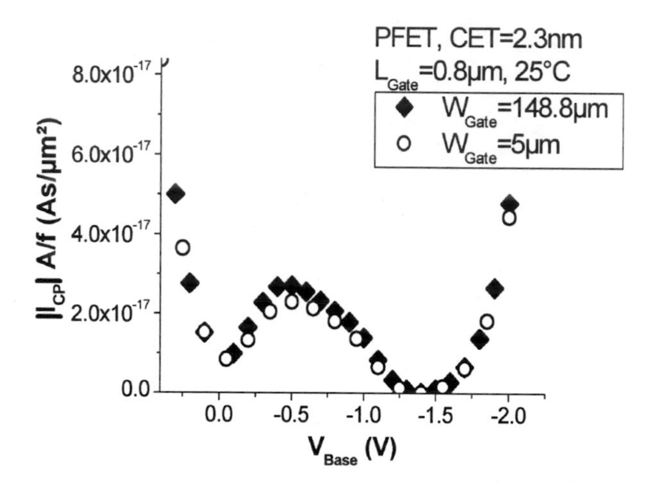

Figure A.5: CP base sweeps at a frequency of 1MHz, 100ns rise- and fall time, and an amplitude of 1.5V for PFETs with a CET of 2.3nm using different test structure geometries. The measured variation is within the die to die difference.

A.5 BSIM4.6.2 Model Card

Model parameters used for the Spice simulations of the gate leakage. Italic parameters are measured electrically, by SIMS and TEM. The bold parameters are adjusted to fit the measured gate leakage. All other values are BSIM4 standard parameters.

BSIM4.6.2 Model Card for PMOS

The BSIM4 model card below was not extracted/obtained from/based on any real technologies. It should not be used for any other purposes except for benchmarking the implementation of BSIM4.6.2 against BSIM Team's standard results using SPICE3f5. Mohan Dunga, Wenwei Yang, Ali Niknejad, and Chenming Hu 05/18/2007.

Model: BSIM4.6.2
Berkeley SPICE3f5 Compatible
Lmin=0.02 Lmax=10 Wmin=0.6 Wmax=20
By Mohan Dunga, Wenwei Yang 05/18/2007

.MODEL P1 PMOS LEVEL = 14

+VERSION = 4.6.1 BINUNIT = 1 PARAMCHK= 1 MOBMOD = 0
+CAPMOD = 2 IGCMOD = 1 IGBMOD = 1 GEOMOD = 1
+DIOMOD = 1 RDSMOD = 0 RBODYMOD= 0 RGATEMOD= 1
+PERMOD = 1 ACNQSMOD= 0 TRNQSMOD= 0 TEMPMOD = 0
+MTRLMOD = 1 CVCHARGEMOD = 0

+EOT = 1.97E-009 VDDEOT = -2.0 ADOS = 1.1 BDOS = 1.0
+TNOM = 85 TOXE = 1.97E-009 TOXP = 3E-009 TOXM = 1.97E-009
+DTOX = -1.03E-009 EPSROX = 15 WINT = 0 LINT = 11E-009
+LL = 0 WL = 0 LLN = 1 WLN = 1
+LW = 0 WW = 0 LWN = 1 WWN = 1
+LWL = 0 WWL = 0 XPART = 0 *TOXREF = 3.0E-9*

+PHIG = 4.4 EPSRGATE = 0

+SAREF = 0 SBREF = 0 WLOD = 0 KU0 = 0
+KVSAT = 0 KVTH0 = 0 TKU0 = 0.0 LLODKU0 = 0
+WLODKU0 = 0 LLODVTH = 0 WLODVTH = 0 LKU0 = 0
+WKU0 = 0 PKU0 = 0.0 LKVTH0 = 0 WKVTH0 = 0
+PKVTH0 = 0.0 STK2 = 0.0 LODK2 = 1.0 STETA0 = 0.0
+LODETA0 = 1.0
+LAMBDA = 0
+VSAT = 3.5E+002
+VTL = 0 XN = 3.0 LC = 0

+RNOIA = 0.577 RNOIB = 0.37
+LINTNOI = 0
+TVOFF = 0.0 TVFBSDOFF = 0.0
+VTH0 = -0.53
+K1 = 0.11 K2 = 0.0 K3 = 0
+K3B = 0 W0 = 0 DVT0 = 0 DVT1 = 0
+DVT2 = 0 DVT0W = 0 DVT1W = 0 DVT2W = 0
+DSUB = 0 MINV = 0 VOFFL = 0 DVTP0 = 0
+MINVCV = 0 VOFFCVL = 0
+DVTP1 = 0 LPE0 = 0 LPEB = 0 *XJ = 8E-008*
+NGATE = 3E020 *NDEP = 1.96E+017 NSD = 1E+021* PHIN = 0
+CDSC = 0.002 CDSCB = 0 CDSCD = 0 CIT = 0
+VOFF = 0.1 NFACTOR = 1.14 ETA0 = 0.0 ETAB = 0
+VFB = 0.48 U0 = 0.0058 UA = 1.0E-009 UB = 1E-019
+UC = 0
+A0 = 1 AGS = 0 +A1 = 0 A2 = 1 B0 = 0 B1 = 0
+KETA = 0.04 DWG = 0 DWB = 0 PCLM = 1
+PDIBLC1 = 0 PDIBLC2 = 0 PDIBLCB = 0 DROUT = 0
+PVAG = 0 DELTA = 0.01 PSCBE1 = 0 PSCBE2 = 1
+FPROUT = 0 PDITS = 0 PDITSD = 0 PDITSL = 0
+RSH = 415 RDSW = 2150 RSW = 2150 RDW = 2150
+RDSWMIN = 0 RDWMIN = 0 RSWMIN = 0 PRWG = 0
+PRWB = 0 WR = 1 ALPHA0 = 0 ALPHA1 = 0
+BETA0 = 0 AGIDL = 0.0002 BGIDL = 2.3E+009 CGIDL = 0
+EGIDL = 0.8 AGISL = 0.0002 BGISL = 2.3E+009 CGISL = 0
+EGISL = 0.8

+AIGBACC = 0.019 BIGBACC = 0.0040 CIGBACC = 0.002
+NIGBACC = 1 **AIGBINV = 0.0183 BIGBINV = 0.065 CIGBINV = 0.002**
+EIGBINV = 1.1 NIGBINV = 3 **AIGC = 0.0135 BIGC = 0.022**
+CIGC = 0.002 AIGS = 0.0118 BIGS = 0.0015 CIGS = 0.002
+NIGC = 1 POXEDGE = 1 NTOX = 1
+AIGD = 0.0118 BIGD = 0.0015 CIGD = 0.002

+XRCRG1 = 12 XRCRG2 = 1
+CGSO = 0 CGDO = 0 CGBO = 0 CGDL = 1.6E-10
+CGSL = 1.6E-10 CKAPPAS = 0.02 CKAPPAD = 0.02 ACDE = 1
+MOIN = 15 NOFF = 1 VOFFCV = 0.02

+KT1 = -0.11 KT1L = 0.0 KT2 = 0 UTE = -1.5
+UA1 = 0 UB1 = 0 UC1 = 0 PRT = 0
+AT = 0

+FNOIMOD = 1 TNOIMOD = 0

+JSS = 0.00092 JSWS = 0.1E-010 JSWGS = 0 NJS = 1
+IJTHSFWD= 0.1 IJTHSREV= 0.1 BVS = 10 XJBVS = 1
+JSD = 0.00092 JSWD = 0.1E-010 JSWGD = 0 NJD = 1
+IJTHDFWD= 0.1 IJTHDREV= 0.1 BVD = 10 XJBVD = 1
+PBS = 1 *CJS = 0.0024* MJS = 0 PBSWS = 1
+CJSWS = 0.34E-010 MJSWS = 0 PBSWGS = 1 CJSWGS = 0
+MJSWGS = 0 PBD = 1 *CJD = 0.0024* MJD = 0
+PBSWD = 1 *CJSWD = 0.34E-010* MJSWD = 0 PBSWGD = 1
+CJSWGD = 0 MJSWGD = 0 TPB = 0 TCJ = 0
+TPBSW = 0 TCJSW = 0 TPBSWG = 0 TCJSWG = 0
+XTIS = 3 XTID = 3

+DMCG = 0E-006 DMCI = 0E-006 DMDG = 0E-006 DMCGT = 0E-007
+DWJ = 0.0E-008 XGW = 0E-007 XGL = 0E-008

+RSHG = 13.3 GBMIN = 1E-020 RBPB = 50 RBPD = 50
+RBPS = 50 RBDB = 50 RBSB = 50 NGCON = 1

+JTSS = 9.2E-4 JTSD = 9.2E-4 JTSSWS = 0.1E-10 JTSSWD = 0.1E-10
+JTSSWGS = 0 JTSSWGD = 0
+NJTS = 20.0 NJTSSW = 20 NJTSSWG = 20 VTSS = 10
+VTSD = 10 VTSSWS = 10 VTSSWD = 10
+NJTSD = 20.0 NJTSSWD = 20 NJTSSWGD = 20
+TNJTS = 0 TNJTSD = 0
+VTSSWGS=10 VTSSWGD=10
+XTSS = 0.02 XTSD = 0.02 XTSSWS = 0.02 XTSSWD = 0.02 XTSSWGS = 0.02
XTSSWGD = 0.02

A.6 TCAD Process Simulation

Excerpts of the program code of the Sentaurus process simulation for a thin gate dielectric PFET. The parameters used for the simulation of the processing are given below.

```
...
# 1. Well Implant
deposit material= Oxide type= isotropic rate= 1.0 time= 0.01

photo mask= PACTIVE thickness= 5.0
implant Phosphorus dose= 5.3e+13 energy= 340.0 tilt= 0.0 rotation=0.0
implant Arsenic dose= 4e+11 energy= 30.0 tilt= 0.0 rotation=0.0
implant Arsenic dose= 1.8e+12 energy=30.0 tilt= 0.0 rotation=0.0
strip Photoresist
...
strip Oxide

# 2. Gate Oxide
deposit material= Oxide type= isotropic rate= 1.0 time= 0.0023
temp_ramp name= tempramp_gate time= 0.18666667 temp= 650.0 ramprate= 22.32
temp_ramp name= tempramp_gate time= 0.2 temp= 900.0 ramprate= 0.0000
temp_ramp name= tempramp_gate time= 0.14 temp= 900.0 ramprate=-29.76
diffuse temp_ramp= tempramp_gate

# 6. Polysilicon
deposit material= Poly type= isotropic rate= 1.0 time= 0.18

temp_ramp name= tempramp_poly time= 0.18666667 temp= 650.0 ramprate= 31.25
temp_ramp name= tempramp_poly time= 0.33333333 temp= 1000.0 ramprate= 0.0000
temp_ramp name= tempramp_poly time= 0.14 temp= 1000.0 ramprate=-41.67
diffuse temp_ramp= tempramp_poly
...
photo mask= POLY thickness= 0.02
etch material= Poly type= anisotropic rate= 0.185 time= 1
etch material= Oxide type= anisotropic rate= 0.003 time= 1
strip Photoresist
...
# 11. Spacer SDE
deposit material= Oxide type= isotropic rate= 1.0 time= 0.017
etch material= Oxide type= anisotropic rate= 0.02 time= 1

# natural through-oxide from cleaning step
deposit material= Oxide type= isotropic rate= 1.0 time= 0.0015
...
```

```
# 12. Implant SDE
photo mask= PACTIVE thickness= 5.0
implant BF2 dose= 4e+14 energy= 3.0 tilt= 0.0 rotation=0
implant Phosphorus dose= 2.0e+12 energy= 40.0 tilt= 28.0 rotation= 0.0
implant Phosphorus dose= 2.0e+12 energy= 40.0 tilt= 28.0 rotation= 90.0
implant Phosphorus dose= 2.0e+12 energy= 40.0 tilt= 28.0 rotation= 270.0
strip Photoresist
...
# 13. Spacer S/D
deposit material= Oxide type= isotropic rate= 1.0 time= 0.033
temp_ramp name= teos time= 22 temp= 630.0 ramprate= 0.0
diffuse temp_ramp= teos
etch material= Oxide type= anisotropic rate= 0.035 time= 1
strip Photoresist

# natural through-oxide from cleaning step
deposit material= Oxide type= isotropic rate= 1.0 time= 0.0015

# 14. S/D Implant
photo mask= PACTIVE thickness= 5.0
implant Boron dose= 2.5e+15 energy= 2.2 tilt= 0.0 rotation=-90.0
strip Photoresist
...
# RTA
temp_ramp name= RTA1 time= 0.033333333 temp= 650.0 ramprate= 185.00
temp_ramp name= RTA1 time= 0.016666667 temp= 1020.0 ramprate= 0.0000
temp_ramp name= RTA1 time= 0.075 temp= 1020.0 ramprate=-82.22
diffuse temp_ramp= RTA1

# RTA
temp_ramp name= RTA2 time= 0.033333333 temp= 650.0 ramprate= 175.00
temp_ramp name= RTA2 time= 0.016666667 temp= 1000.0 ramprate= 0.0000
temp_ramp name= RTA2 time= 0.075 temp= 1000.0 ramprate=-77.78
diffuse temp_ramp= RTA2
...
# 15. Contacts
etch material= Oxide type= anisotropic rate= 0.007 time=1

# Al metalization
deposit material = Aluminum type = isotropic rate = 1.0 time= 0.007
photo mask= METAL thickness= 0.02
etch material= Aluminum type= anisotropic rate= 0.1 time= 1 isotropic.overetch= 0.05
strip Photoresist
...
```

```
# 6. Constant Polysilicon Doping
select z= -2e+20 name= NetActive PolySilicon store
...

# Contact definition instead of Al metal
contact name= "gate" box PolySilicon adjacent.material= Gas
ylo= [expr -1*$Yg] yhi= $Yg
xlo= [expr $Xtop-0.1] xhi= [expr $Xtop+0.1]
contact bottom name=substrate Silicon
contact point x= [expr $XtopSi-5e-3] y= [expr $Ymax-5e-3]
name= "drain" Aluminum
contact point x= [expr $XtopSi-5e-3] y= [expr -$Ymax+5e-3]
name= "source" Aluminum

struct smesh= n@node@
exit
```

A.7 TCAD Device Simulation $I_{Drain}V_{Gate}$

The program code of the Sentaurus device simulation for a PFET. The parameters used for the simulation are given in the table.

```
!(
set SIGN -1.0
set DG "hQuantumPotential"
set cTemp "hTemperature"
set EQN0 "Poisson hQuantumPotential Electron Hole"
set EQNS "Poisson hQuantumPotential Electron Hole hTemperature Temperature"

)!

File
* input files:
Grid= "@tdr@"
Parameter="@parameter@"
* output files:
Plot= "@tdrdat@"
Current="@plot@"
Output= "@log@"

Electrode
Name="source" Voltage= 0.0 Resistor= 450
Name="drain" Voltage= 0.0 Resistor= 450
Name="gate" Voltage= 0.0 Barrier=-0.288
Name="substrate" Voltage= 0.0

Physics
Temperature=358
Hydrodynamic( !(puts $cTemp)! )
EffectiveIntrinsicDensity( OldSlotboom )

Physics(Material="Silicon")
Temperature=358
!(puts $DG)!
Mobility(
PhuMob
```

```
HighFieldSaturation
Enormal
)


Solve
*- Creating initial guess:
Coupled(Iterations= 100 LineSearchDamping= 1e-4) Poisson !(puts $DG)!
Coupled !(puts $EQN0)!
Coupled !(puts $EQNS)!

*- Ramp to drain to Vd
Quasistationary(
InitialStep= 1e-2 Increment= 1.35
MinStep= 1e-5 MaxStep= 0.2
Goal Name="drain" Voltage=!(puts [expr $SIGN*@Vdlin@])!
) Coupled !(puts $EQNS)!

*- Vg sweep
NewCurrentFile="IdVg "
Quasistationary(
DoZero
InitialStep= 1e-3 Increment= 1.5
MinStep= 1e-5 MaxStep= 0.04
Goal Name="gate" Voltage=!(puts [expr $SIGN*2])!
) Coupled !(puts $EQNS)!
CurrentPlot( Time=(Range=(0 1) Intervals= 30) )
```

Table A.4: Model Parameters of the Sentaurus Device Simulation

Parameter	Value (electron,hole)	Model
mumax_ As	1.4170e+03cm^2/Vs	Philips Unified Mobility
mumin_ As	52.2cm^2/Vs	Philips Unified Mobility
theta_ As	2.285	Philips Unified Mobility
n_ ref_ As	9.68e+16cm^{-3}	Philips Unified Mobility
alpha_ As	0.68	Philips Unified Mobility
mumax_ P	1.4140e+03cm^2/Vs	Philips Unified Mobility
mumin_ P	68.5cm^2/Vs	Philips Unified Mobility
theta_ P	2.285	Philips Unified Mobility
n_ ref_ P	9.20e+16cm^{-3}	Philips Unified Mobility
alpha_ P	0.711	Philips Unified Mobility
mumax_ B	1.47050e+02cm^2/Vs	Philips Unified Mobility
mumin_ B	44.9cm^2/Vs	Philips Unified Mobility
theta_ B	2.247	Philips Unified Mobility
n_ ref_ B	2.23e+17cm^{-3}	Philips Unified Mobility
alpha_ B	0.719	Philips Unified Mobility
nref_ D	4e+20cm^{-3}	Philips Unified Mobility
nref_ A	7.2e+20cm^{-3}	Philips Unified Mobility
cref_ D	0.21	Philips Unified Mobility
cref_ A	0.5	Philips Unified Mobility
me_ over_ m0	1	Philips Unified Mobility
mh_ over_ m0	1.285	Philips Unified Mobility
f_ CW	2.459, 2.459	Philips Unified Mobility
f_ BH	3.828, 3.828	Philips Unified Mobility
f_ gf	1, 1	Philips Unified Mobility
f_ scr	0, 0 [cm^5]/Vs	Philips Unified Mobility
B	4.75E+07, 9.9250E+06 [cm/s]	Enormal
C	5.8E+02, 2.947E+03 [cm$^{5/3}$/V$^{2/3}$s]	Enormal
N0	1, 1 [cm^{-3}]	Enormal
lambda	0.125, 0.0317	Enormal
k	1, 1	Enormal
delta	5.820e+14, 2.0546e+14 [V/s]	Enormal
A	2, 2	Enormal
alpha	0, 0	Enormal
N1	1, 1 [cm^{-3}]	Enormal
nu	1, 1	Enormal
eta	5.82e+30, 2.0546e+30 [V^2/s·cm]	Enormal
l_ crit	1e-06, 1e-06 [cm]	Enormal
beta0	1.109, 1.213	HighFieldDependence
betaexp	0.66, 0.17	HighFieldDependence
K_ dT	0.2, 0.2	HighFieldDependence
E0_ TrEf	4e+03, 4e+03	HighFieldDependence

Parameter	Value (electron,hole)	Model
Ksmooth_ TrEf	1, 1	HighFieldDependence
vsat0	1.07e+07, 8.37e+06	HighFieldDependence
vsatexp	0.87, 0.52	HighFieldDependence
gamma	3.6, 5.6	QuantumPotentialParameters
theta	0.5, 0.5	QuantumPotentialParameters
xi	1, 1	QuantumPotentialParameters
eta	1, 1	QuantumPotentialParameters
tau_ w_ ele	0.3ps	EnergyRelaxationTime
tau_ w_ hol	0.25ps	EnergyRelaxationTime

A.8 Influence of Model Parameters on Current Simulations

Current generation by band to band tunneling becomes important at higher voltages. Band to band tunneling vanishes near insulator interfaces as no states in the energy gap of the dielectric are available [54]. To account for the effect, a potential difference of 1.1V, and a distance of $1{\cdot}10^{-30}$cm where no tunneling occurs is introduced.

The field enhanced trap assisted current generation is simulated with the Hurkx model. In the Hurkx model, the effective mass is a fit parameter. For simplicity equal fit parameters for electron and holes are assumed [38]. The effective mass parameter effects the slope of the leakage current

A fixed or uniform trap energy distribution is introduced. The trap energy determines the temperature dependence of the leakage current curves (Fig. A.6). The slope of leakage current is independent from the trap energy. The trap energy used in the simulation is fitted to the measured activation energy (equation 2.2). The simulations show that the activation energy decreases with trap energy closer to mid gap (Fig. A.6(b)). The trap density is proportional to the amount of leakage current.

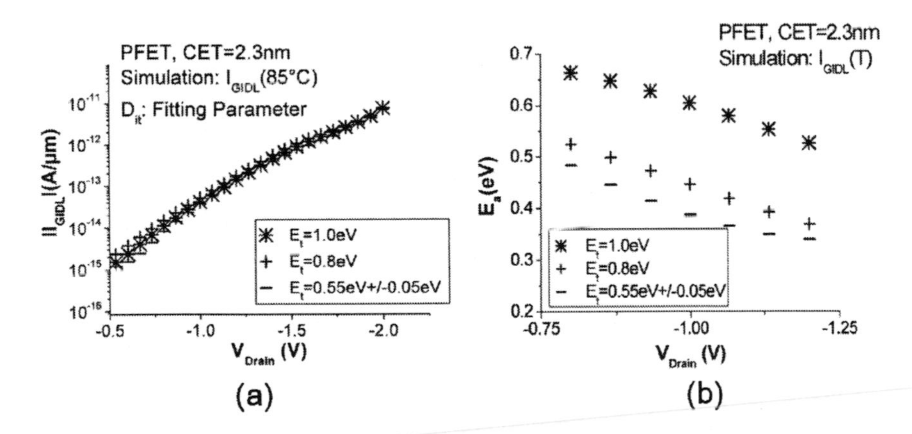

Figure A.6: Simulated trap assisted GIDL characteristics using different energy levels of the defects.
(a) Simulated GIDL versus drain bias. The uniform interface trap density is adjusted to reach similar currents, and compare the leakage currents slopes for the PFETs with a CET below 3nm.
(b) Activation energy calculated from the temperature dependence of the simulated GIDL curves.

The fixed mobility parameter influences the flattening of the interface trap assisted leakage current with increasing bias (Fig. A.7). At high voltages, and low mobilities the current is reduced due to a too slow transport of the carriers away from the interface. But the leakage current in this bias region is controlled by band to band tunneling.

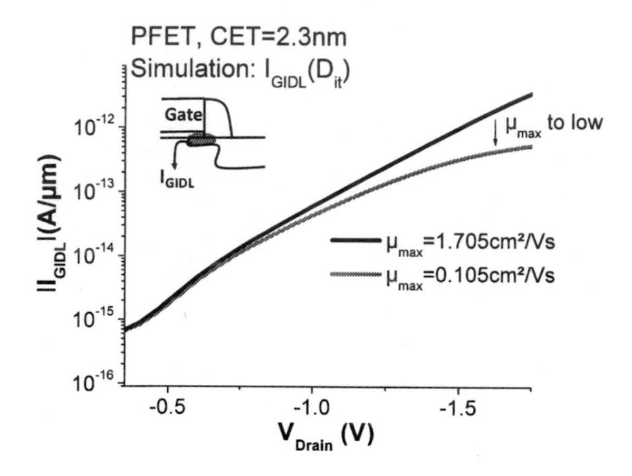

Figure A.7: Change of simulated interface trap assisted current with constant mobility parameter for PFETs with a CET below 3nm.

B Appendix: Devices with Different Junction Implant

B.1 Processing: Devices with Junction Implant

Table B.1: Sample Description for PFETs with Different Carbon Implants

Sample		$V_{th}2$,C↑,Halo↔	$V_{th}2$,C↓,Halo↔
Implant			
Well	P	340keV, $5.3 \cdot 10^{13}$atm/cm^2	340keV, $5.3 \cdot 10^{13}$atm/cm^2
	As	30keV, $4 \cdot 10^{11}$atm/cm^2	30keV, $4 \cdot 10^{11}$atm/cm^2
V_{th} (CET=2.3nm)	As	30keV, $1.8 \cdot 10^{12}$atm/cm^2	30keV, $1.8 \cdot 10^{12}$atm/cm^2
V_{th} (CET=5.2nm)	As	30keV, $1 \cdot 10^{12}$atm/cm^2	30keV, $1 \cdot 10^{12}$atm/cm^2
Poly	B	2.5keV,$6 \cdot 10^{15}$atm/cm^2	2.5keV,$6 \cdot 10^{15}$atm/cm^2
SDE	Ge	20keV, $3 \cdot 10^{14}$atm/cm^2	20keV, $3 \cdot 10^{14}$atm/cm^2
	BF_2	3keV, $4 \cdot 10^{14}$atm/cm^2	3keV, $4 \cdot 10^{14}$atm/cm^2
	C	**4keV, $4 \cdot 10^{14}$atm/cm^2**	**4keV, $3.5 \cdot 10^{14}$atm/cm^2**
Halo	P	40keV, $2.3 \cdot 10^{13}$atm/cm^2	40keV, $2.3 \cdot 10^{13}$atm/cm^2
S/D	Ge	20keV, $3 \cdot 10^{14}$atm/cm^2	20keV, $3 \cdot 10^{14}$atm/cm^2
	B	2.2keV, $2.5 \cdot 10^{15}$atm/cm^2	2.2keV, $2.5 \cdot 10^{15}$atm/cm^2

Table B.2: Sample Description for PFETs with Different V_{th} Implants

Sample		$\mathbf{V}_{th}\mathbf{1}$,C↑,Halo↔	$\mathbf{V}_{th}\mathbf{2}$,C↑,Halo↔
Implant			
Well	P	340keV, $5.3{\cdot}10^{13}$atm/cm^2	340keV, $5.3{\cdot}10^{13}$atm/cm^2
	As	30keV, $4{\cdot}10^{11}$atm/cm^2	30keV, $4{\cdot}10^{11}$atm/cm^2
V_{th} (CET=2.3nm)	**As**	**30keV, $\mathbf{1.5{\cdot}10^{12}}$atm/cm^2**	**30keV, $\mathbf{1.8{\cdot}10^{12}}$atm/cm^2**
V_{th} (CET=5.2nm)	**As**	**30keV, $\mathbf{1.1{\cdot}10^{12}}$atm/cm^2**	**30keV, $\mathbf{1{\cdot}10^{12}}$atm/cm^2**
Poly	B	2.5keV,$6{\cdot}10^{15}$atm/cm^2	2.5keV,$6{\cdot}10^{15}$atm/cm^2
SDE	Ge	20keV, $3{\cdot}10^{14}$atm/cm^2	20keV, $3{\cdot}10^{14}$atm/cm^2
	BF_2	3keV, $4{\cdot}10^{14}$atm/cm^2	3keV, $4{\cdot}10^{14}$atm/cm^2
	C	4keV, $4{\cdot}10^{14}$atm/cm^2	4keV, $4{\cdot}10^{14}$atm/cm^2
Halo	P	40keV, $2.3{\cdot}10^{13}$atm/cm^2	40keV, $2.3{\cdot}10^{13}$atm/cm^2
S/D	Ge	20keV, $3{\cdot}10^{14}$atm/cm^2	20keV, $3{\cdot}10^{14}$atm/cm^2
	B	2.2keV, $2.5{\cdot}10^{15}$atm/cm^2	2.2keV, $2.5{\cdot}10^{15}$atm/cm^2

Table B.3: Sample Description for PFETs with Different Halo Implants

Sample		V_{th}2,C↓,**Halo↔**	V_{th}2,C↓,**Halo↓**	V_{th}2,C↓,**Halo↑**
Implant				
Well	P	340keV, $5.3 \cdot 10^{13}$atm/cm^2	340keV, $5.3 \cdot 10^{13}$atm/cm^2	340keV, $5.3 \cdot 10^{13}$atm/cm^2
	As	30keV, $4 \cdot 10^{11}$atm/cm^2	30keV, $4 \cdot 10^{11}$atm/cm^2	30keV, $4 \cdot 10^{11}$atm/cm^2
V_{th} CET=2.3nm	As	30keV, $1.8 \cdot 10^{12}$atm/cm^2	30keV, $1.8 \cdot 10^{12}$atm/cm^2	30keV, $1.8 \cdot 10^{12}$atm/cm^2
V_{th} CET=5.2nm	As	30keV, $1 \cdot 10^{12}$atm/cm^2	30keV, $1 \cdot 10^{12}$atm/cm^2	30keV, $1 \cdot 10^{12}$atm/cm^2
Poly	B	2.5keV,$6 \cdot 10^{15}$atm/cm^2	2.5keV,$6 \cdot 10^{15}$atm/cm^2	2.5keV,$6 \cdot 10^{15}$atm/cm^2
SDE	Ge	20keV, $3 \cdot 10^{14}$atm/cm^2	20keV, $3 \cdot 10^{14}$atm/cm^2	20keV, $3 \cdot 10^{14}$atm/cm^2
	BF_2	3keV, $4 \cdot 10^{14}$atm/cm^2	3keV, $4 \cdot 10^{14}$atm/cm^2	3keV, $4 \cdot 10^{14}$atm/cm^2
	C	4keV, $3.5 \cdot 10^{14}$atm/cm^2	4keV, $3.5 \cdot 10^{14}$atm/cm^2	4keV, $3.5 \cdot 10^{14}$atm/cm^2
Halo	**P**	**40keV, $2.3 \cdot 10^{13}$atm/cm^2**	**40keV, $2.1 \cdot 10^{13}$atm/cm^2**	**40keV, $2.5 \cdot 10^{13}$atm/cm^2**
S/D	Ge	20keV, $3 \cdot 10^{14}$atm/cm^2	20keV, $3 \cdot 10^{14}$atm/cm^2	20keV, $3 \cdot 10^{14}$atm/cm^2
	B	2.2keV, $2.5 \cdot 10^{15}$atm/cm^2	2.2keV, $2.5 \cdot 10^{15}$atm/cm^2	2.2keV, $2.5 \cdot 10^{15}$atm/cm^2

B.2 NFET Interface Defect Characterization

Different NFET samples used to compare the interface defect concentration dependent on the oxide deposition process. The NFETs are processed similar to the PFET samples. Only the doping implantation steps are varied (table B.4). The NFETs are analyzed to exclude the boron diffusion effects that occurs in PFET devices due to polysilicon doping.

Table B.4: Sample Description for NFETs

Sample		NFET1	NFET2
Implant			
Well	B	150keV, $5.3{\cdot}10^{13}$atm/cm^2	150keV, $5.3{\cdot}10^{13}$atm/cm^2
	In	135keV, $1{\cdot}10^{13}$atm/cm^2	135keV, $1{\cdot}10^{13}$atm/cm^2
	B	10keV, $6{\cdot}10^{11}$atm/cm^2	10keV, $6{\cdot}10^{11}$atm/cm^2
V_{th} (CET=2.5nm)	B	**10keV, $2.3{\cdot}10^{13}$atm/cm^2**	**10keV, $1.8{\cdot}10^{13}$atm/cm^2**
V_{th} (CET=5.3nm)	B	**10keV, $5{\cdot}10^{12}$atm/cm^2**	**10keV, $4{\cdot}10^{12}$atm/cm^2**
Poly	P	7keV,$5{\cdot}10^{15}$atm/cm^2	7keV,$5{\cdot}10^{15}$atm/cm^2
SDE (CET=2.5nm)	As	2keV, $1.2{\cdot}10^{15}$atm/cm^2	2keV, $1.2{\cdot}10^{15}$atm/cm^2
SDE (CET=5.3nm)	As	11keV, $1.2{\cdot}10^{14}$atm/cm^2	11keV, $1.2{\cdot}10^{14}$atm/cm^2
	P	8keV, $1{\cdot}10^{13}$atm/cm^2	8keV, $1{\cdot}10^{13}$atm/cm^2
Halo (CET=2.5nm)	B	10keV, $1.6{\cdot}10^{13}$atm/cm^2	10keV, $1.6{\cdot}10^{13}$atm/cm^2
Halo (CET=5.3nm)	B	10keV, $1.8{\cdot}10^{13}$atm/cm^2	10keV, $1.8{\cdot}10^{13}$atm/cm^2
S/D	As	12keV, $3{\cdot}10^{15}$atm/cm^2	12keV, $3{\cdot}10^{15}$atm/cm^2

B.3 Off-Current of the PFETs

In this section the current of the different leakage paths on the voltage, and the device geometry are compared.

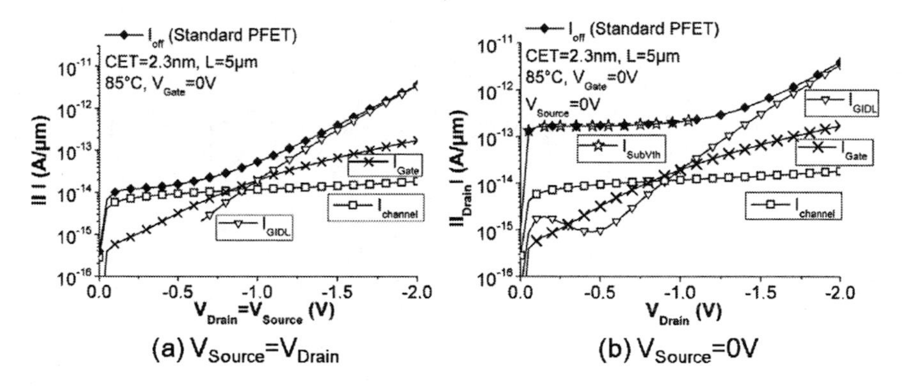

Figure B.1: Leakage current contributions depending on the voltage for a 5μm long thin gate dielectric PFET. Different bias conditions are applied at the source:
(a) V_{Source}=V_{Drain}.
(b) V_{Source}=0V.

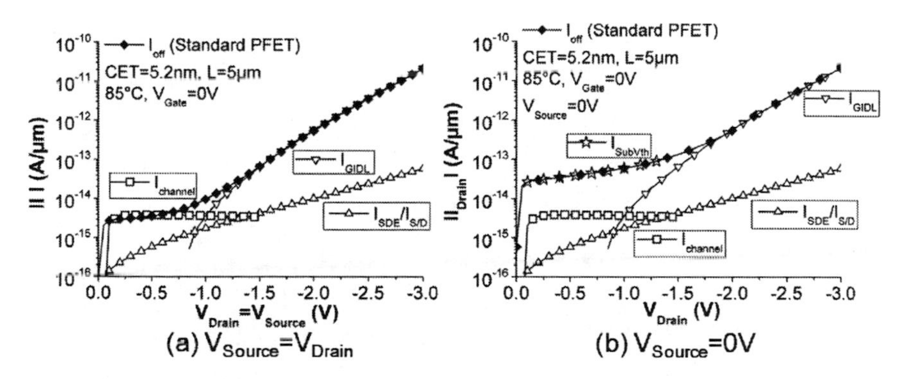

Figure B.2: Leakage current contributions depending on the voltage for a 5μm long thick gate dielectric PFET. Different bias conditions are applied at the source:
(a) V_{Source}=V_{Drain}.
(b) V_{Source}=0V.

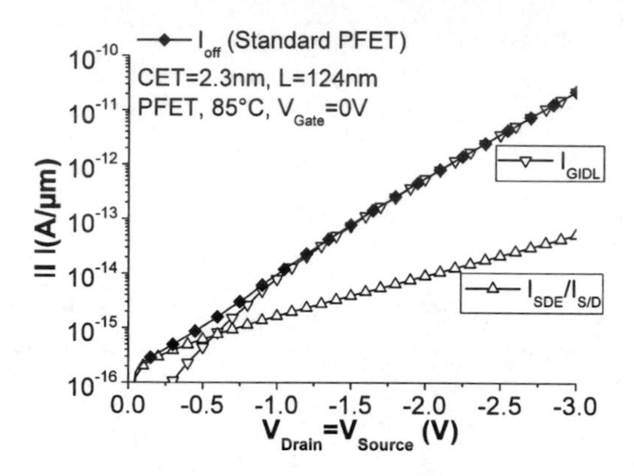

Figure B.3: Leakage current contributions depending on the voltage for a 124nm long thick gate dielectric PFET.

B.4 Model Parameters of the Sentaurus Device Simulation

Equal parameters are assigned for electrons, and holes assuming one charge carrier being dominant in a certain leakage mechanism (table B.5). The barrier, and resistor parameters are chosen accordingly to the best fit of the subthreshold current characteristics (table B.6). Different trap regions are defined in the simulations. They are summarized in table B.7. An example command file for the Sentaurus device simulations is also given. The examples contains S/D diode leakage current calculations.

Table B.5: Model Parameters of the Sentaurus Device Simulation

Parameter	Value	Model
μ^1_{max}	$1.705 cm^2/Vs$	Constant Mobility
Exponent	2.2	Constant Mobility
μ_{tunnel}	$0.05 cm^2/Vs$	Constant Mobility
E_{g0}	-0.01595eV	OldSlotboom
E_{bgn}	0.009eV	OldSlotboom
N_{ref}	$1 \cdot 10^{-17} cm^{-3}$	OldSlotboom
σ^1	$1 \cdot 10^{-15} cm^2$	Traps
N: EnergyMid	0eV from Midgap	Traps (Level)
D_{it}: EnergyMid	0eV	Traps (uniform)
D_{it}: EnergySigma	0.5eV	Traps (uniform)
$m^{*\,1}_{bulk}$	$0.12 m_{electron}$	Hurkx
$m^{*\,1}_{it}$	$0.6 m_{electron}$	Hurkx
A^1	$1 \cdot 10^{14} cm^{-3} s^{-1}$	Nonlocal Tunneling (Path 1)
B	$1 \cdot 10^7 Vcm^{-1}$	Nonlocal Tunneling (Path 1)
D	0.037eV	Nonlocal Tunneling (Path 1)
R	0	Nonlocal Tunneling (Path 1)
$dPot^1$	1.1V	Nonlocal Tunneling (Path 1)
dDist	$1 \cdot 10^{-30} cm$	Nonlocal Tunneling (Path 1)

[1]Parameters changed compared to the standard parameter file

Table B.6: Barrier and Resistor parameters for the different PFETs

Device	Resistor	Barrier
S/D diode	450Ω	/
PFET CET=2.3nm L_{Gate}=65nm	450Ω	-0.288V
PFET CET=2.3nm L_{Gate}=5μm	10000Ω	-0.288V
PFET CET=5.2nm L_{Gate}=5μm	10000Ω	-0.288V

Table B.7: Trapping Parameters of the Sentaurus Device Simulation for PFET with a C Dose of 4.0·10^{14}atm/cm^2

Trap Region	Charge Trapping Efficiency	Place
$\sigma \cdot N_{back}$	$5.2 \cdot 10^{-4} cm^{-1}$	SDE S/D N_{back} n-well
$\sigma \cdot N_I$	$1.2 \cdot 10^{-0} cm^{-1}$ (CET=5.2nm) $1.9 \cdot 10^{-1} cm^{-1}$ (CET=2.3nm)	D_{itedge} D_{it} N_I $N_{Isurface}$ N_{back}
$\sigma \cdot N_{Isurface}$	$0.042 \cdot \sigma \cdot N_I$	
$\sigma \cdot D_{it}$	$3.5 \cdot 10^{-7} eV^{-1}$ $1.85 \cdot 10^{-6} eV^{-1}$ (CET=2.3nm)	
$\sigma \cdot D_{itedge}$	$1.3 \cdot 10^{-4} eV^{-1}$ $2 \cdot 10^{-5} eV^{-1}$ (CET=2.3nm)	

σ - capture cross section (cm^2)
N - trap density in the silicon (cm^{-3})
D_{it} - trap density per energy at the interface (cm^{-2}eV^{-1})

Junction.cmd

```
!(
set SIGN -1.0
set DG "hQuantumPotential"
set cTemp "hTemperature"
set EQN0 "Poisson hQuantumPotential Electron Hole"
set EQNS "Poisson hQuantumPotential Electron Hole")!
```

```
File
* input files:
Grid= "@tdr@"
Parameter="@parameter@"
* output files:
Plot= "@tdrdat@"
Current="@plot@"
Output= "@log@"

Electrode
Name="top" Voltage= 0.0 Resistor= 450
Name="substrate" Voltage= 0.0

Physics Temperature=348
EffectiveIntrinsicDensity( OldSlotboom )

Physics(Material="Silicon")
Temperature=348
!(puts $DG)!
Mobility(ConstantMobility)
Traps (
(hNeutral Conc=5.2e11 Level EnergyMid=0 fromMidBandGap Tunneling(Hurkx)))


Insert= "PlotSection_des.cmd"
Insert= "MathSection_des.cmd"

Solve
*- Creating initial guess:
Coupled(Iterations= 100 LineSearchDamping= 1e-4) Poisson !(puts $DG)!
Coupled  !(puts $EQN0)!
Coupled  !(puts $EQNS)!

*- Vd sweep
NewCurrentFile="Junction"
Quasistationary(
DoZero
InitialStep= 1e-3 Increment= 1.5
MinStep= 1e-5 MaxStep= 0.04
Goal  Name="top" Voltage=!(puts [expr -5])!
) Coupled  !(puts $EQNS)!
CurrentPlot( Time=(Range=(0 1) Intervals= 30) )
```

B.5 Leakage Current Depending on Oxide Residuals prior to Implantation

During the processing, it was suspected that possibly oxide residuals remain for the thick gate dielectric PFET prior to SDE implantation (Fig. B.4). Those kind of residuals would change the implantation profiles, and effect the leakage current. The effect of a 2nm thick oxide residual is simulated. The leakage current are increased due to the oxide residual. The increase is caused by the reduction of SDE junction depth, and a higher halo concentration at the junction (Fig. B.5). The influence of the remaining oxide is approximately a factor of 2. So even an oxide residual with 2nm thickness can not explain the increased trap density for the thick gate dielectric device estimated by the simulation.

Figure B.4: Variation of the Sentaurus process simulation to analyze the effect of remaining oxide residuals

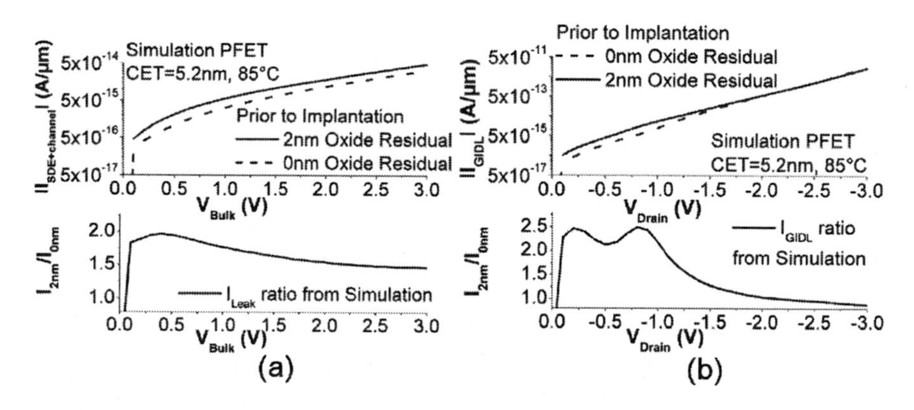

Figure B.5: Simulated ,and measured leakage current changes due to oxide residuals. The oxide residuals change the doping profiles, and so effect the leakage.
(a) Simulated SDE- plus channel leakage current are presented.
(b) Simulated GIDL current is shown.

B.6 Gate Induced Drain Leakage Statistic Variations

The figure shows the statistic variation of about 10% using five GIDL current measurements on different dies.

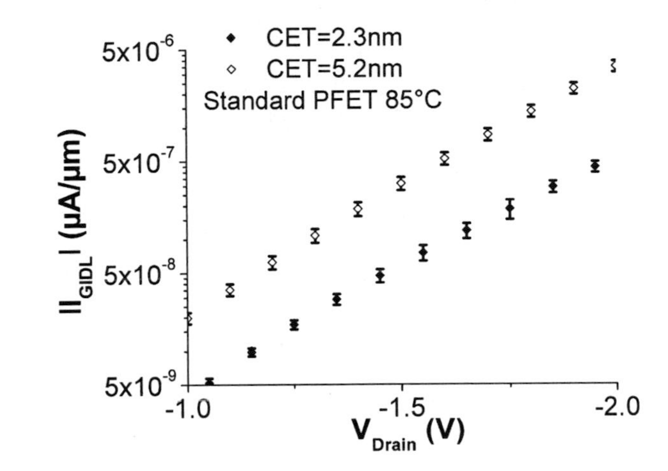

Figure B.6: Measured GIDL current over drain voltage, and the error bars of the measurements for PFETs at 85°C.

B.7 Gate Leakage

Dependence of gate leakage for a 5μm long transistor on the gate voltage. An increased importance of the gate to bulk current in the depletion regime can be observed. Figure B.7 gives the Fowler Nordheim plots for the gate leakage of the PFET devices. The gate oxide field is determined using the split CV measurements using a ϵ_r of 3.9 (section 3.2). Under inversion a linear behavior can be found. The slope is with $1.23 \cdot 10^9$V/m very low. The barrier extracted from this value is 0.43eV. The hole mass of the HVB is assumed to be 0.41 of the electron mass [40]. The barrier height is expected to be between 1.9eV and 4.5eV [132]. But the Fowler Nordheim approximation does not take depletion layers in the silicon, barrier lowering due to image force, and quantum mechanical effects into account.

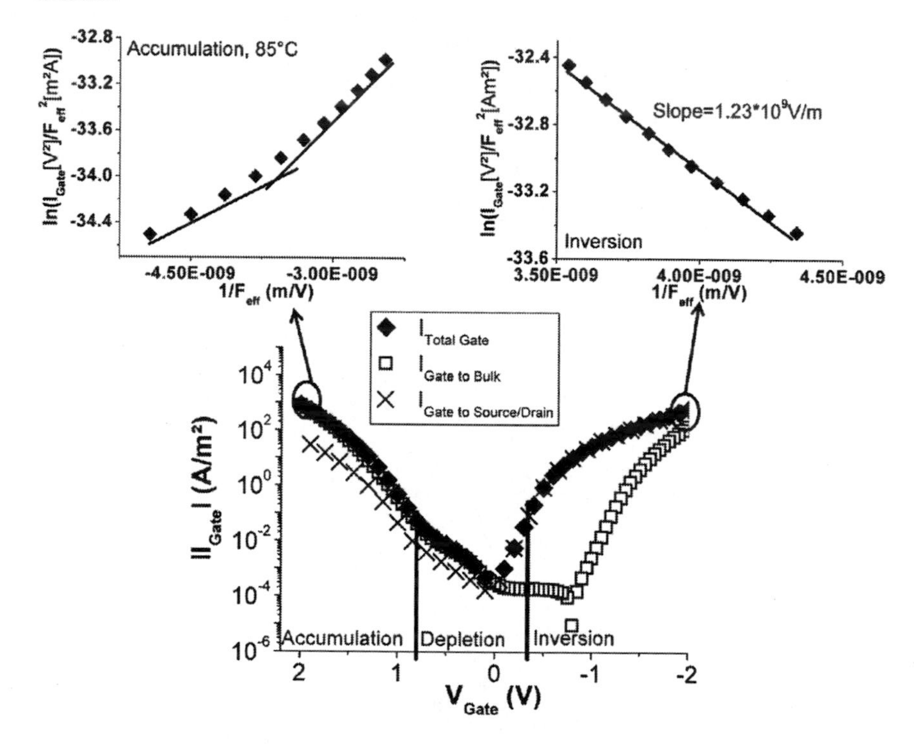

Figure B.7: Gate leakage in dependence on the voltage for a 5μm long PFET with standard implants, and a CET of 2.3nm. The Fowler Nordheim plots are deduced from the measurement (upper part). Fowler Nordheim behavior is assumed under inversion with a slope of $1.23 \cdot 10^9$V/m.

B.8 Effect of Carbon on S/D Junction Profile

The SIMS profile shown here is taken from a different experiment with similar processing conditions. A germanium amorphization energy of 20keV, and dose of $3{\cdot}10^{14}$ atm/cm^2 is chosen. The boron implantation conditions in the S/D junction are 2.3keV, and $2.5{\cdot}10^{15}$ atm/cm^2. A device with a 3keV, and $7{\cdot}10^{14}$ atm/cm^2 carbon implant is compared with a sample with no carbon. It can clearly be seen that the carbon implantation changes the S/D junction profile at the metallurgical junction.

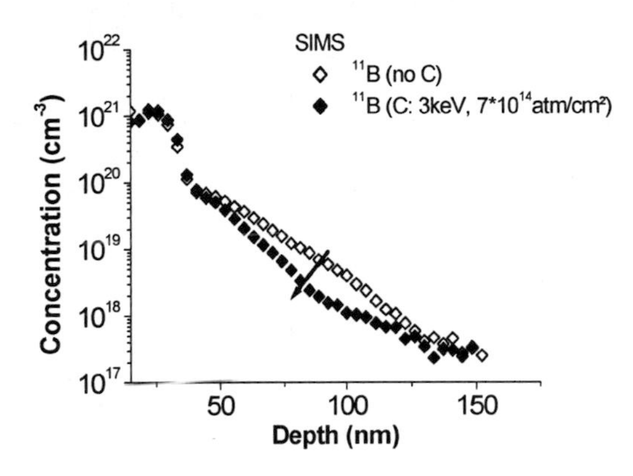

Figure B.8: The effect of a carbon co-implantation on the S/D junction profile is shown for a different experiment [85]. The boron doping profile above the metallurgical junction is changed by the carbon implantation.

B.9 Change of Junction Capacitance with Doping Profiles

Devices of a different sample series measured with, and without DRAM anneal. Clearly the change of junction capacitance with doping profile can be observed (Fig. B.9).

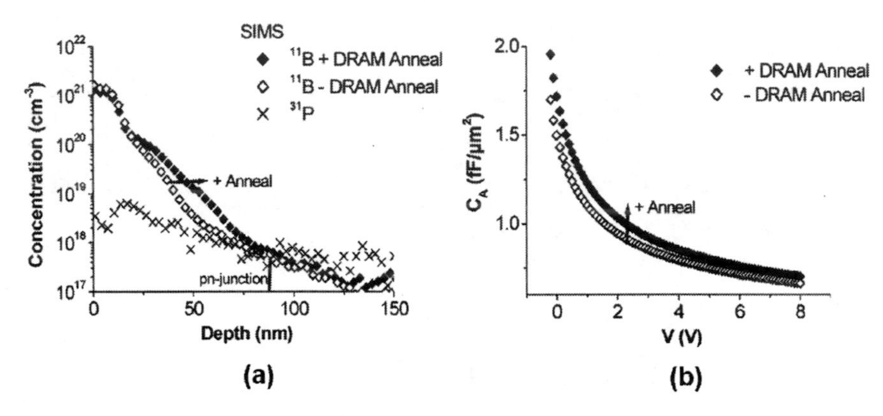

Figure B.9: SIMS analysis, and CV measurement at the samples with, and without DRAM anneal.
(a) Increased boron diffusion due to higher thermal budget measured by SIMS analysis [85].
(b) Increased junction capacitance measured by CV for PFETs with DRAM anneal.

B.10 Basic Electrical Characteristics of the PFETs

Table B.8: Basic Characteristics of PFETs (CET≈2nm, L_{Gate}=5μm)

	V_{th}1,C↑,Halo↔	V_{th}2,C↑,Halo↔	V_{th}2,C↓,Halo↓	V_{th}2,C↓,Halo↔	V_{th}2,C↓,Halo↑
CET (nm)	2.33±0.01	2.32±0.02	2.32±0.01	2.32±0.01	2.32±0.01
L_{ov} (nm)	7.2±0.3	6.6±0.2	6.7±1.2	7.0±0.4	6.7±0.6
μ_{max} (nm)	44.33±3.78	43.58±3.05	43.74±0.96	42.59±3.26	42.64±0.84
V_{th} (V) V_{Drain}=-0.05V (CV*)	-0.35±0.05	-0.4±0.05	-0.4±0.05	-0.4±0.05	-0.4±0.05
V_{th} (V) V_{Drain}=-0.05V (IV)	-0.47±0.01	-0.52±0.01	-0.52±0.02	-0.52±0.01	-0.51±0.01
V_{th} (V) V_{Drain}=-1V (IV)	-0.48±0.05	-0.51±0.04	-0.51±0.04	-0.51±0.03	-0.50±0.03
V_{fb} (V) (CV*)	0.8±0.05	0.8±0.05	0.8±0.05	0.8±0.05	0.81±0.07
I_{on} (μA/μm)	1.86±0.31	1.52±0.42	1.52±0.43	1.49±0.33	1.47±0.35
S_{Vth} (mV/dec) V_{Drain}=-0.05V	85.9±0.4	88.9±0.7	89.0±1.0	88.6±0.9	89.0±0.3

All values are given for measurements on 5 dies (wafer center).

*V_{fb} and V_{th} are taken from the point of maximum, and minimum slope of the CV characteristic.

Table B.9: Basic Characteristics of PFETs (CET≈5nm, L_{Gate}=5μm)

	V_{th}1,**C**↑,Halo↔	V_{th}2,**C**↑,Halo↔	V_{th}2,**C**↓,Halo↓	V_{th}2,**C**↓,Halo↔	V_{th}2,**C**↓,Halo↑
CET (nm)	5.21±0.02	5.22±0.01	5.23±0.03	5.23±0.02	5.19±0.02
L_{ov} (nm)	7.7±0.7	7.3±0.5	7.9±0.6	7.3±0.4	5.0±0.9
μ_{max} (nm)	52.4±1.7	52.0±0.6	52.1±1.7	51.5±2.0	51.9±2.5
V_{th} (V) V_{Drain}=-0.05V (CV*)	-0.45±0.05	-0.5±0.05	-0.5±0.05	-0.51±0.07	-0.51±0.07
V_{th} (V) V_{Drain}=-0.05V (IV)	-0.59±0.01	-0.60±0.01	-0.59±0.01	-0.60±0.01	-0.59±0.01
V_{th} (V) V_{Drain}=-1V (IV)	-0.56±0.04	-0.56±0.01	-0.56±0.01	-0.56±0.01	-0.56±0.01
V_{fb} (V) (CV*)	0.8±0.05	0.8±0.05	0.8±0.05	0.79±0.07	0.81±0.07
I_{on} (μA/μm)	0.15±0.05	0.49±0.03	0.10±0.01	0.50±0.04	0.1±0.01
S_{Vth} (mV/dec) V_{Drain}=-0.05V	97.0±3.0	102.5±0.9	102.9±1.1	102.1±1.6	102.2±1.3

All values are given for measurements on 5 dies (wafer center).
*V_{fb} and V_{th} are taken from the point of maximum, and minimum slope of the CV characteristic.

B.11 The SDE Leakage Dependent on V_{th}- and Halo Implants

The SDE leakage is increased with increasing halo implant (Fig. B.10). The measurements done on the 65nm long devices show a high variation in SDE current. So it is not clear if the V_{th} dose variation for the thin gate dielectric device is high enough to effect the SDE leakage current.

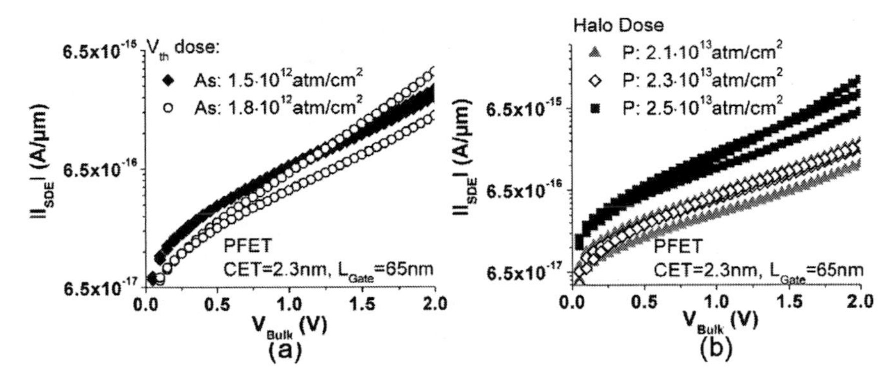

Figure B.10: SDE leakage current for PFETs with a CET of 2.3nm, and a channel length of 65nm at 85°C. Different implant dose variations are compared:
(a) V_{th} implant.
(b) Halo implant.

B.12 Interface Trap Density Dependent on the V_{th} Implant

Average interface trap density dependent on oxide type, and V_{th} implant. No change in interface trap density within the measurement variation is found (Fig. B.11). The change in channel leakage for the thick gate dielectric devices can not be explained by an increase in interface trap density.

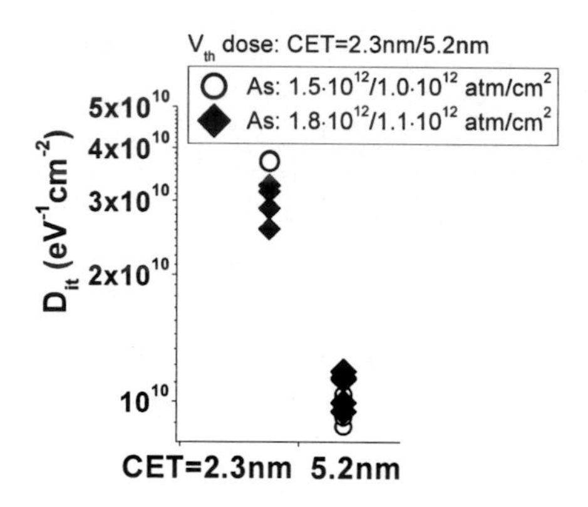

Figure B.11: Average PFET interface trap density dependent on oxide type, and V_{th} implant. No change in interface trap density within the measurement variation is found.

B.13 Subthreshold Current Dependent on V_{th}- and Halo Implant for PFETs

The subthreshold current for the long channel PFET with a CET of 5.2nm is independent of the halo implant (Fig. B.12). The minor variation of the V_{th} implantation dose do not significantly increase the subthreshold current within the measurement error.

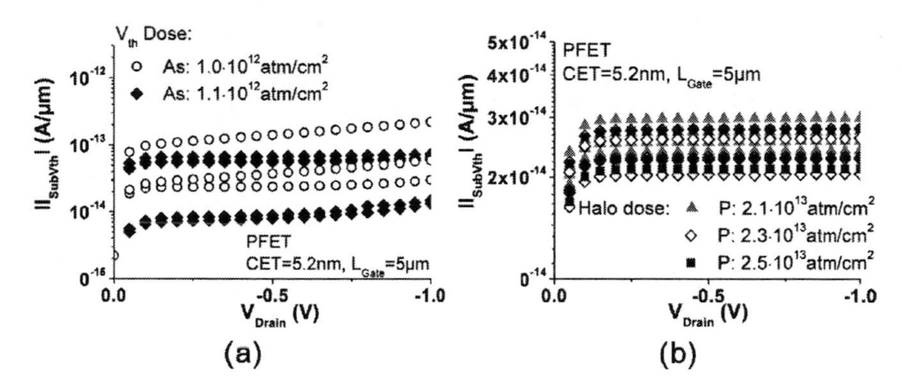

Figure B.12: Subthreshold current at 85°C for PFETs with a CET of 5.2nm, and a channel length of 5μm. Different implant dose variations are compared:
(a) V_{th} implant.
(b) Halo implant.

C Appendix: Devices with High-k Oxide

C.1 Processing: High-k Devices

Table C.1: Implantation Energies and Doses for the High-k Devices with SiO_2 and SiN Spacer

Implant	Species	PFET
Well	P	340keV, $5.3 \cdot 10^{13}$atm/cm^2
V_{th}	As (CET≈2nm)	30keV, $9 \cdot 10^{11}$atm/cm^2
V_{th}	As (CET≈5.5nm)	30keV, $5 \cdot 10^{11}$atm/cm^2
Poly	P	7keV, $5 \cdot 10^{15}$atm/cm^2
SDE	Ge	15keV, $3 \cdot 10^{14}$atm/cm^2
	BF_2	3keV, $3 \cdot 10^{14}$atm/cm^2
Halo	P	40keV, $1.8 \cdot 10^{13}$atm/cm^2
S/D	Ge	15keV, $3 \cdot 10^{14}$atm/cm^2
	BF_2	10keV, $2.5 \cdot 10^{15}$atm/cm^2

Table C.2: Implantation Energies and Doses for the High-k Devices with Different High-k Etches

Implant	Species	PFET	Species	NFET
Well	P	340keV, $5.3 \cdot 10^{13}$atm/cm^2	B	150keV, $5.3 \cdot 10^{13}$atm/cm^2
V_{th}	As (CET≈2nm)	30keV, $9 \cdot 10^{11}$atm/cm^2	B (CET≈2nm)	10keV, $7 \cdot 10^{12}$atm/cm^2
V_{th}	As (CET≈5.5nm)	30keV, $5 \cdot 10^{11}$atm/cm^2	B (CET≈5.5nm)	10keV, $1.8 \cdot 10^{12}$atm/cm^2
Poly	P	7keV, $5 \cdot 10^{15}$atm/cm^2	P	7keV, $5 \cdot 10^{15}$atm/cm^2
SDE	Ge	15keV, $3 \cdot 10^{14}$atm/cm^2		
	BF_2	3keV, $3 \cdot 10^{14}$atm/cm^2	As (CET≈2nm)	3keV, $1 \cdot 10^{15}$atm/cm^2
			P (CET≈5.5nm)	8keV, $2.5 \cdot 10^{13}$atm/cm^2
			As (CET≈5.5nm)	8keV, $1.2 \cdot 10^{14}$atm/cm^2
Halo	P	40keV, $1.8 \cdot 10^{13}$atm/cm^2	BF_2 (CET≈2nm)	45keV, $1.4 \cdot 10^{13}$atm/cm^2
			BF_2 (CET≈5.5nm)	60keV, $1.5 \cdot 10^{13}$atm/cm^2
S/D	Ge	15keV, $3 \cdot 10^{14}$atm/cm^2		
	BF_2	10keV, $2.5 \cdot 10^{15}$atm/cm^2	As	16keV, $3 \cdot 10^{15}$atm/cm^2

Table C.3: Implantation Energies and Doses for the High-k Devices with SiO_2+SiN SDE Spacer

Implant	Species	PFET	Species	NFET
Well	P	340keV, $5.3 \cdot 10^{13}$atm/cm^2	B	150keV, $5.3 \cdot 10^{13}$atm/cm^2
V_{th}	As (CET≈2nm)	30keV, $9 \cdot 10^{11}$atm/cm^2	B (CET≈2nm)	10keV, $7 \cdot 10^{12}$atm/cm^2
V_{th}	As (CET≈5.5nm)	30keV, $5 \cdot 10^{11}$atm/cm^2	B (CET≈5.5nm)	10keV, $1.8 \cdot 10^{12}$atm/cm^2
Poly	P	7keV, $5 \cdot 10^{15}$atm/cm^2	P	7keV, $5 \cdot 10^{15}$atm/cm^2
SDE	Ge	15keV, $3 \cdot 10^{14}$atm/cm^2		
	BF_2	3keV, $3 \cdot 10^{14}$atm/cm^2	As (CET≈2nm)	3keV, $1 \cdot 10^{15}$atm/cm^2
			P (CET≈5.5nm)	8keV, $2.5 \cdot 10^{13}$atm/cm^2
			As (CET≈5.5nm)	8keV, $1.2 \cdot 10^{14}$atm/cm^2
Halo	P	40keV, $1.6 \cdot 10^{13}$atm/cm^2	BF_2 (CET≈2nm)	45keV, $1.4 \cdot 10^{13}$atm/cm^2
			BF_2 (CET≈5.5nm)	60keV, $1.7 \cdot 10^{13}$atm/cm^2
S/D	Ge	15keV, $3 \cdot 10^{14}$atm/cm^2		
	BF_2	10keV, $2.5 \cdot 10^{15}$atm/cm^2	As	16keV, $3 \cdot 10^{15}$atm/cm^2

Table C.4: Implantation Energies and Doses for the SiON Reference Device (Processing: section 4.1)

Implant	Species	PFET	Species	NFET
Well	P	340keV, $5.3 \cdot 10^{13}$atm/cm^2	B	150keV, $5.3 \cdot 10^{13}$atm/cm^2
	As	30keV, $1.4 \cdot 10^{12}$atm/cm^2	B	10keV, $8.0 \cdot 10^{11}$atm/cm^2
V_{th}	As (CET=2.5nm)	30keV, $2.0 \cdot 10^{12}$atm/cm^2	B (CET=2.7nm)	10keV, $1.6 \cdot 10^{13}$atm/cm^2
V_{th}	As (CET=5.1nm)	30keV, $5 \cdot 10^{11}$atm/cm^2	B (CET=5.3nm)	10keV, $2.0 \cdot 10^{12}$atm/cm^2
Poly	B	2.5keV, $6 \cdot 10^{15}$atm/cm^2	P	7keV, $5 \cdot 10^{15}$atm/cm^2
SDE	Ge	15keV, $3 \cdot 10^{14}$atm/cm^2		
	BF_2	3keV, $3 \cdot 10^{14}$atm/cm^2	As (CET=2.7nm)	3keV, $1 \cdot 10^{15}$atm/cm^2
			P (CET=5.3nm)	8keV, $2.5 \cdot 10^{13}$atm/cm^2
			As (CET=5.3nm)	8keV, $1.2 \cdot 10^{14}$atm/cm^2
Halo	P	40keV, $2.3 \cdot 10^{13}$atm/cm^2	BF_2 (CET=2.7nm)	45keV, $1.6 \cdot 10^{13}$atm/cm^2
			BF_2 (CET=5.3nm)	45keV, $1.6 \cdot 10^{13}$atm/cm^2
S/D	Ge	20keV, $3 \cdot 10^{14}$atm/cm^2		
	B	2.2keV, $2.5 \cdot 10^{15}$atm/cm^2	As	12keV, $3 \cdot 10^{15}$atm/cm^2

C.2 Transistor Off-Currents

This section gives an overview of the intrinsic MOSFET off-state leakage currents measured in the different test structures, and experiments.

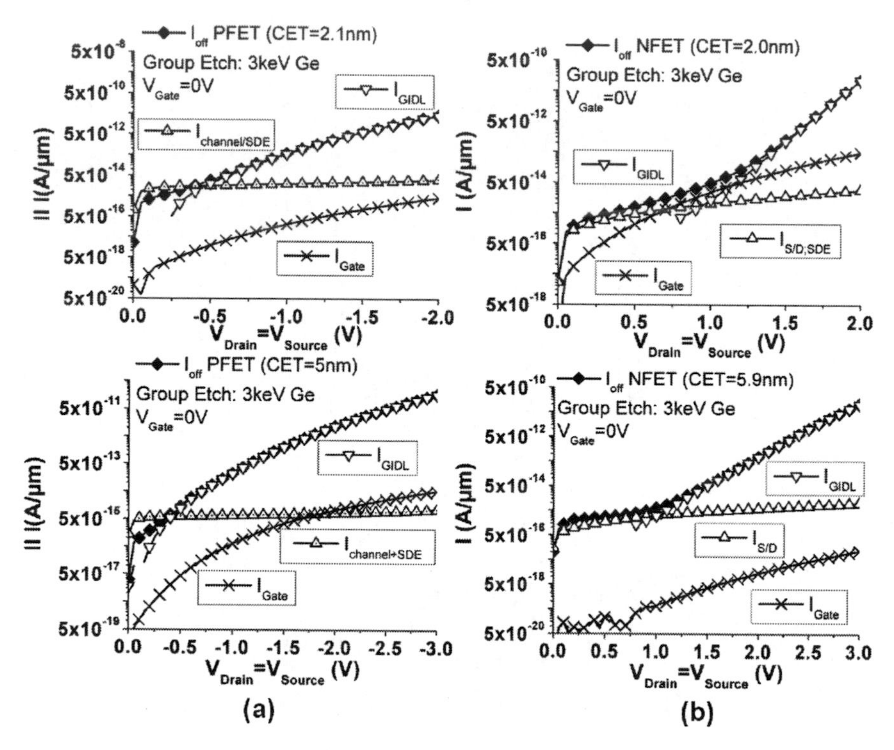

Figure C.1: Leakage current in off-state for V_{Source}=V_{Drain} for transistors with an 3keV Ge etch at 85°C are shown. Current voltage plots for the 800nm long thin gate dielectric devices are displayed in the upper part of the figure, and for the thick gate dielectric transistors in the lower part of the picture.
(a) PFET
(b) NFET

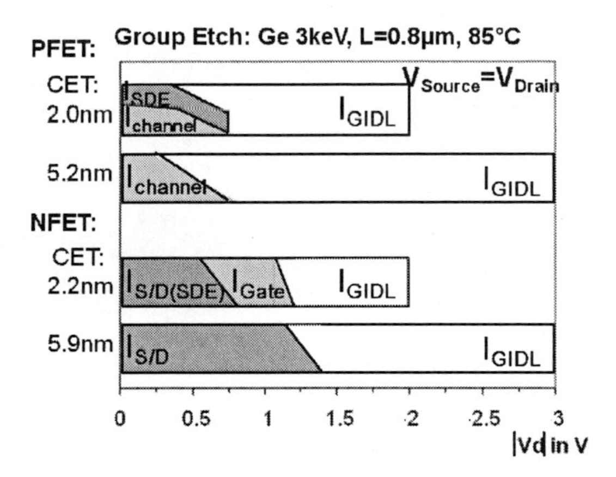

Figure C.2: The dominant leakage current contributions to the total transistors leakage depending on the voltage for long channel transistors with thick, and thin gate dielectric at 85°C is shown in the bar graph.

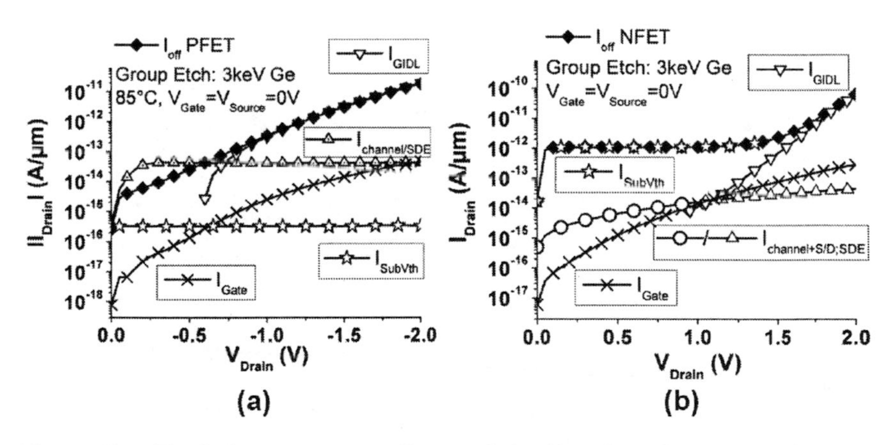

Figure C.3: The leakage current in off-state of the 10μm long devices with thin gate dielectric from the etch sample set with a 3keV Ge implant at 85°C is shown for a:
(a) PFET.
(b) NFET.

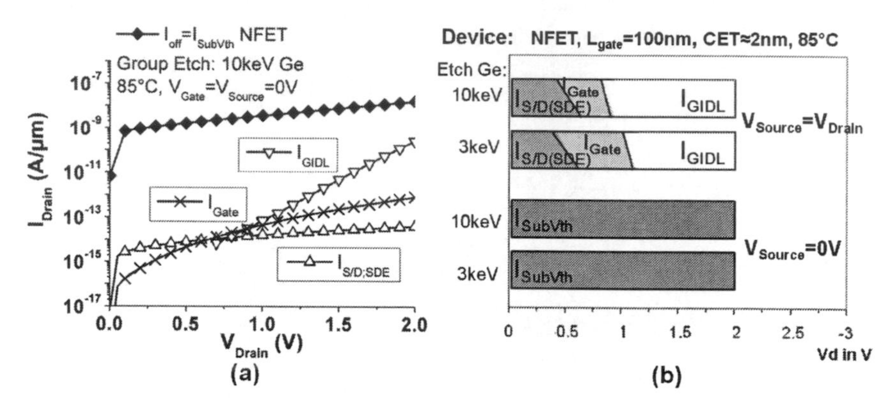

Figure C.4: NFET leakage current in off-state for V_{Source}=0V for a typical device with L_{Gate}=100nm, and a CET of about 2nm at 85°C is shown.
(a) Current voltage plots of the NFET with 10keV germanium implant energy is shown.
(b) The main leakage current contribution depending on the voltage for the different Ge etch energies is shown in the bar graph.

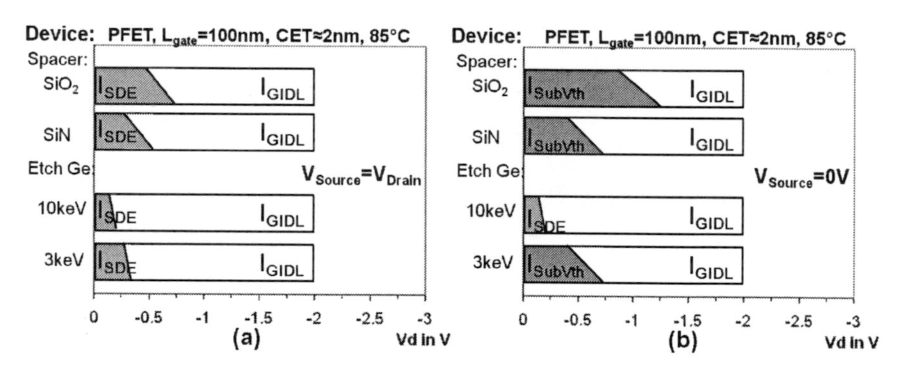

Figure C.5: The main off-current contribution depending on the voltage, for all investigated high-k PFETs with L_{Gate}=100nm, and a CET of about 2nm, is shown in the bar graph. Different bias conditions are applied at the source:
(a) V_{Source}=V_{Drain}.
(b) V_{Source}=0V.

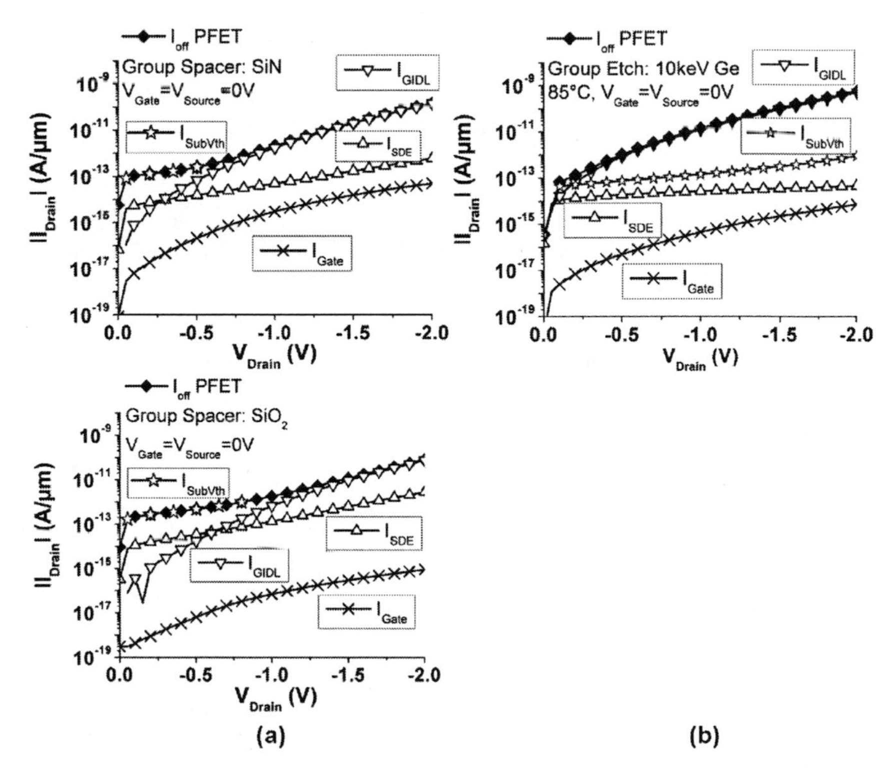

Figure C.6: Typical PFET leakage current in off-state for V_{Source}=0V for 100nm long devices with a thin gate dielectric at 85°C.
(a) PFETs from the sample set with different SDE spacers are presented: SiN spacer in the upper part, and SiO_2 spacer in the lower part of the graph.
(b) A transistor with 10keV germanium implant from the sample set with different etch energies is shown.

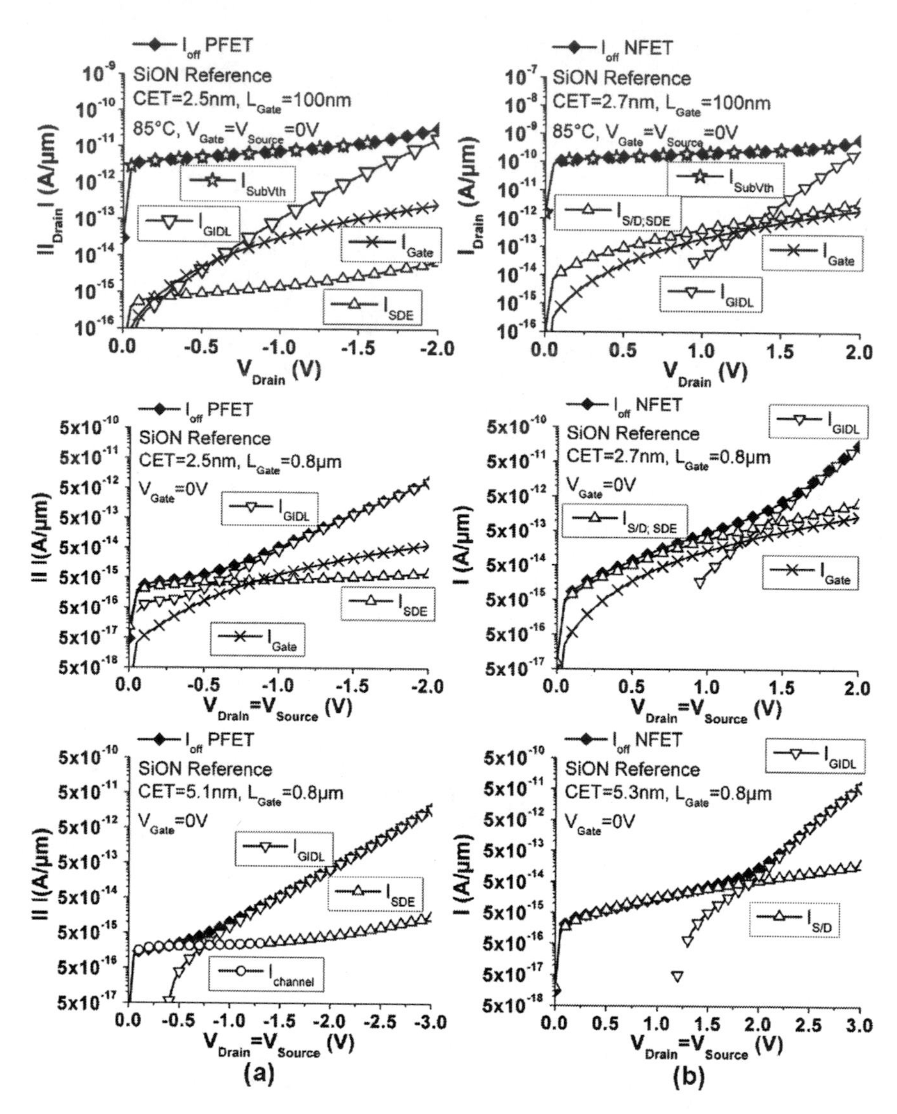

Figure C.7: Leakage current in off-state of a SiON reference sample for 100nm and 800nm long transistors at 85°C are shown:
(a) PFET.
(b) NFET.

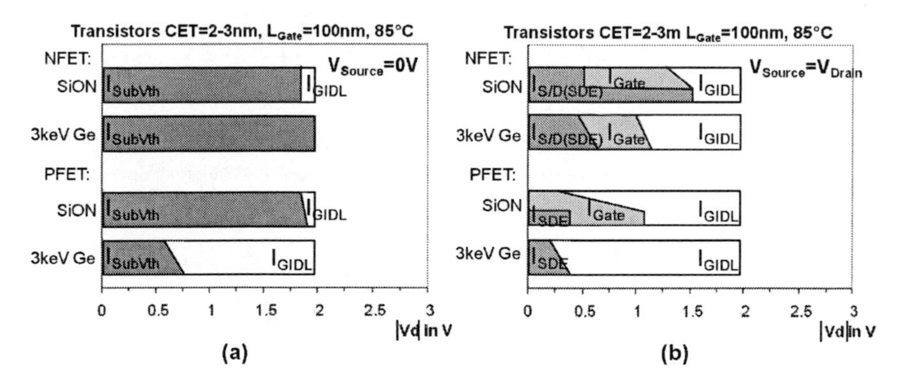

Figure C.8: The main off-current contribution depending on the voltage at 85°C for PFET, and NFET with L_{Gate}=100nm, and a CET between 2nm and 3nm are shown. SiON reference samples are compared to the transistors with high-k gate stack of the sample set with different etches (3keV Ge). Different bias conditions are applied at the source:
(a) V_{Source}=0V.
(a) V_{Source}=V_{Drain}.

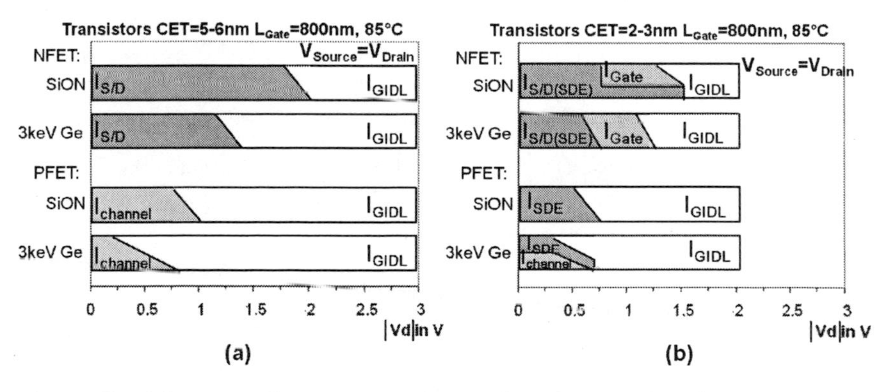

Figure C.9: The main off-current contribution depending on the voltage at 85°C for PFET, and NFET with L_{Gate}=800nm are shown. SiON reference samples are compared to the transistors with high-k gate stack of the sample set with different etches (3keV Ge). Transistors with different gate dielectrics are compared:
(a) CET=5-6nm.
(b) CET=2-3nm.

C.3 Analytical Approximation of the Diode Current

The fit parameter of the analytical approximation [43] of the measured diode leakage are given in table C.5. The Hurkx mechanism is the main leakage current mechanism below ±3V [43] (Fig. C.10).

Table C.5: Fitting Factors of the Diode Current

Sample	V_{int} (V)	r	m*	B_{BTBT} (A/V)	E_t (eV)	*N·σ (cm^{-1})
	F_{eff}		I_{Hurkx}, I_{BTBT}	I_{BTBT}	I_{Hurkx}, I_{Gen}	
PFET						
Ge Etch Set						
3keV Ge	0.79	0.47	0.25	$1\cdot 10^8$	0.56	0.0043
10keV Ge	0.78	0.47	0.25	$4\cdot 10^8$	0.56	0.022
Spacer Set	0.81	0.5	0.25	-	0.56	0.015-0.062
SiON	0.8	0.45	0.25	-	0.58	0.0048
NFET						
Ge Etch Set						
3keV Ge	1.1	0.5	0.3	$3\cdot 10^7$	0.5	0.03
10keV Ge	1.1	0.5	0.3	$5\cdot 10^7$	0.5	0.17
SiON	1.07	0.5	0.3	-	0.56	0.12

* Estimated for an thermal velocity of 230000m/s.
Median measured curve is fitted for the SiON reference-, and high-k Ge etch sample set. Highest, and lowest diode leakage is fitted for the high-k spacer sample set because of the high statistic variation of this sample.

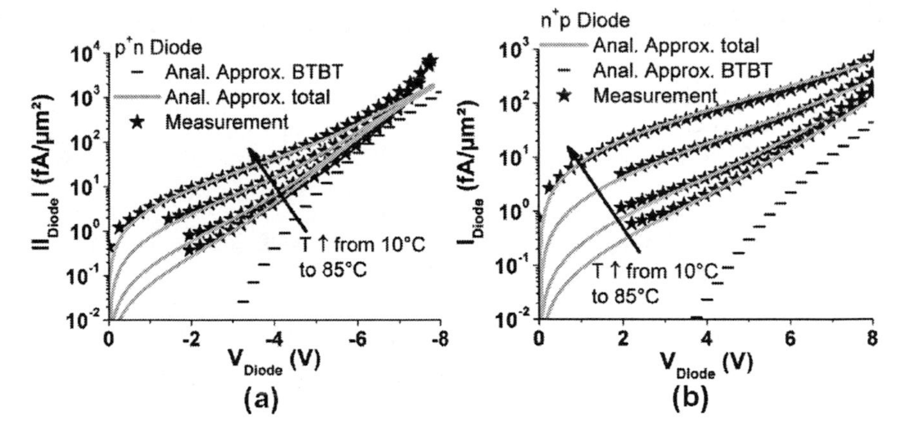

Figure C.10: Comparision of measurement, and analytical approximation [43] of the diode leakage depending on the bias. The current contribution of the BTBT component is approximated as explained in [43]. Diodes of the etch set with a 10keV germanium implant are compared.
(a) p^+n diodes
(b) n^+p diodes

C.4 PFET Model Parameters of the Sentaurus Device Simulation

Equal parameters are assigned for electron, and holes assuming one charge carrier being dominant in a certain leakage mechanism (table C.6). The barrier, and resistor parameters are chosen accordingly to the best fit of the subthreshold current characteristics (table C.7). Changes in the barrier parameter are consistent with changes in the fixed charges of the gate dielectric. Different trap regions are defined in the simulations. They are summarized in tables C.8, and C.9.

Table C.6: Model Parameters of the Sentaurus Device Simulation

Parameter	Value	Model
μ_{max}^1	5.3cm^2/Vs	Constant Mobility
Exponent	2.2	Constant Mobility
μ_{tunnel}	0.05cm^2/Vs	Constant Mobility
E_{g0}	-0.01595eV	OldSlotboom
E_{bgn}	0.009eV	OldSlotboom
N_{ref}	$1{\cdot}10^{-17}$cm^{-3}	OldSlotboom
σ^1	$1{\cdot}10^{-15}$cm^2	Traps
N: EnergyMid	0eV from Midgap	Traps (Level)
N_{it}: EnergyMid	0.1eV from Midgap	Traps (Level)
$m_{bulk}^{*\ 1}$	$0.08m_{electron}$	Hurkx
$m_{it}^{*\ 1}$	0.5$m_{electron}$	Hurkx
A^1	$1{\cdot}10^{14}$cm^{-3}s^{-1}	Nonlocal Tunneling (Path 1)
B	$1{\cdot}10^7$Vcm^{-1}	Nonlocal Tunneling (Path 1)
D	0.037eV	Nonlocal Tunneling (Path 1)
R	0	Nonlocal Tunneling (Path 1)
dPot[1]	1.1V	Nonlocal Tunneling (Path 1)
dDist	$1{\cdot}10^{-30}cm$	Nonlocal Tunneling (Path 1)

[1]Parameters changed compared to the standard parameter file

Table C.7: PFET Barrier and Resistor parameters (CET≈2nm)

Sample	Device	Resistor	Barrier	CET	SDE Spacer Thickness
Set Spacer: SiO_2	S/D diode	100Ω	/	/	/
	PFET L_{Gate}=100nm	100Ω	0.45V	2.3nm	8nm
	PFET L_{Gate}=10μm	100Ω	0.57V	2.3nm	8nm
Group Spacer: SiN	PFET L_{Gate}=100nm	100Ω	0.45V	2.0nm	11nm
Group Etch	S/D diode	100Ω	/	/	/
	PFET L_{Gate}=100nm	100Ω	0.45V	2.0nm	11nm
	PFET L_{Gate}=10μm	100Ω	0.57V	2.0nm	11nm

Table C.8: Trapping Parameters of the Sentaurus Device Simulation for PFETs with Different Spacers (CET≈2nm)

Trap Region	Charge Trapping Efficiency	Place
$\sigma \cdot N_{back}$	$9.2 \cdot 10^{-3} cm^{-1}$	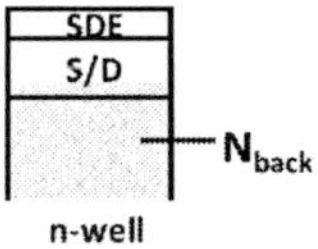
$\sigma \cdot N_I$	$4 \cdot 10^{-0} cm^{-1}$	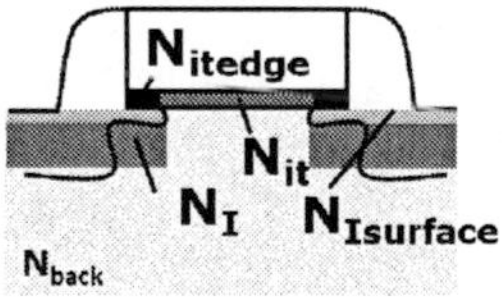
$\sigma \cdot N_{Isurface}$	$0.01 \cdot \sigma \cdot N_I$	
$\sigma \cdot N_{it}$	$1.8 \cdot 10^{-5} cm^{-1}$	
$\sigma \cdot N_{itedge}$	$2.4 \cdot 10^{-4} cm^{-1}$	

σ - capture cross section (cm^2)

N - trap density in the silicon (cm^{-3})

Table C.9: Trapping Parameters of the Sentaurus Device Simulation for PFET with Different Ge Etches (CET≈2nm)

Trap Region	Charge Trapping Efficiency	Place
$\sigma \cdot N_{back}$	$1.16 \cdot 10^{-3} cm^{-1}$ (3keV Ge) $5.80 \cdot 10^{-3} cm^{-1}$ (10keV Ge)	SDE S/D N_{back} n-well
$\sigma \cdot N_I$	$1 \cdot 10^{-2} cm^{-1}$ (3keV Ge) $4 \cdot 10^{-2} cm^{-1}$ (10keV Ge)	N_{itedge} N_{it} N_I $N_{Isurface}$ N_{back}
$\sigma \cdot N_{Isurface}$	$0.01 \cdot \sigma \cdot N_I$	
$\sigma \cdot N_{it}$	$8.75 \cdot 10^{-6} cm^{-1}$	
$\sigma \cdot N_{itedge}$	$1 \cdot 10^{-4} cm^{-1}$(3keV Ge) $1.65 \cdot 10^{-3} cm^{-1}$(10keV Ge)	

σ - capture cross section (cm^2)
N - trap density in the silicon (cm^{-3})

Temperature Dependence of Simulated and Measured PFET Leakages

A trap energy of 0.65eV is chosen for the interface traps to fit the temperature dependence of the PFET channel leakage, and GIDL (Fig. C.11, C.12). The temperature dependence of the GIDL current for PFETs with SiO_2 spacer in the model slightly exceeds (Fig. C.12). For the bulk silicon defects, a mid gap trap is chosen. The measured, and simulated temperature dependence of the diode leakage current is in good agreement (Fig. C.11). For the leakage currents from the traps N_I, the temperature dependence is poorly reproduced (Fig. C.13). For the perimeter leakage current from the surface traps ($N_{Isurface}$), and interface edge traps (N_{itedge}), that occurs when the bulk is biased, a good agreement is reached (Fig. C.13).

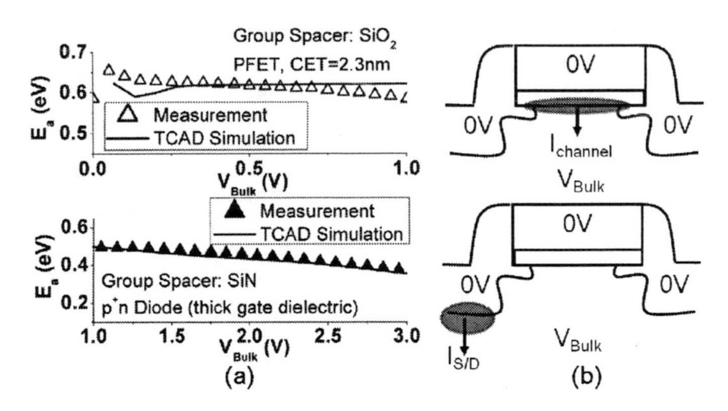

Figure C.11: The temperature dependence of the channel leakage for a thin gate dielectric PFET is shown in the upper part. The activation energy of a p^+n-S/D diode is presented in the lower part.
(a) Measured, and simulated activation energy when the bulk is biased.
(b) Leakage paths of the channel-, and the S/D diode current are shown.

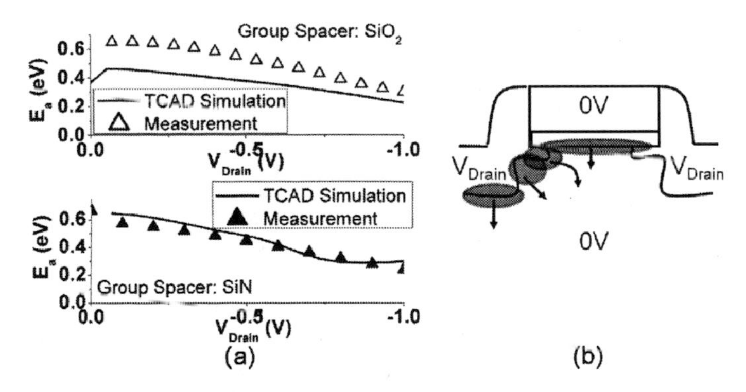

Figure C.12: Temperature dependence of the drain current (mainly GIDL current) shown for thin gate dielectric PFETs from the sample set with different spacers (L_{Gate}=100nm). A good agreement for the simulated, and measured activation energy is reached for the transistors with SiN spacer (lower part of the figure). The E_a of the model exceeds the measurements for the sample with SiO_2 spacer (upper part of the figure).
(a) Measured, and simulated activation energy of the drain current
(b) Sketch of the bias condition, and the leakage path of measured current, is shown.

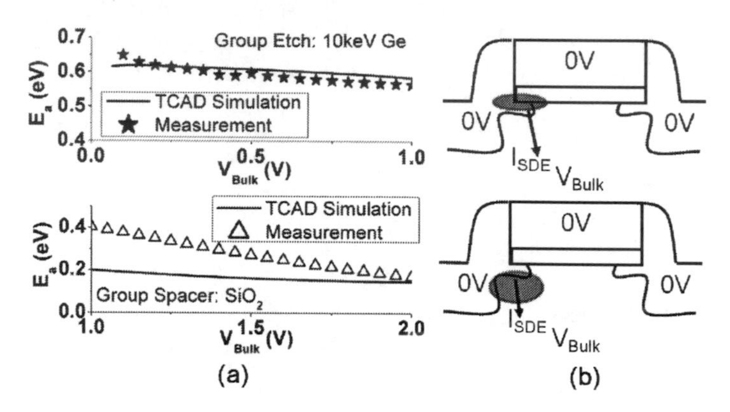

Figure C.13: Temperature dependence of the source/drain extension leakage current shown for thin gate dielectric PFETs with L_{Gate}=100nm. The upper part of the figure presents a typical device of the etch sample set with a 10keV germanium implant. The simulation accurately reproduce the activation energy of the traps at the gate oxide to silicon interface. The lower part of the figure presents a typical PFET of the spacer sample set. The simulated temperature dependence of the silicon bulk traps agrees only poorly with the measurements.
(a) Measured, and simulated activation energy of the SDE current when the bulk is biased.
(b) Leakage path of the source drain extension current is shown.

C.5 Hurkx Traps Assisted Tunneling: TCAD vs. Analytical Approximation

The Hurkx trap assisted tunneling model decreases the capture cross section of the generation current by the tunneling component (B_{Hurkx}).

$$\sigma_{Hurkx} = \frac{\sigma_{e/h}}{1 + B_{Hurkx}} \tag{C.1}$$

The tunneling component factor (B_{Hurkx}) is derived from the the ratio of emission probability with, and without tunneling [38].

For the analytical approximation, only the maximum effective field at the junction is taken into account. B_{Hurkx} simplifies to [43]:

$$\begin{aligned} B_{Hurkx} &= 2\sqrt{3\pi}\frac{F_{eff}}{F_\Gamma}exp\left[\frac{F_{eff}}{F_\Gamma}\right]^2 \\ F_\Gamma &= \frac{\sqrt{24m^* \cdot (kT)^3}}{e\hbar} \end{aligned} \tag{C.2}$$

For the TCAD simulation, the used approximation takes the effective electric field at every grid point of the 2D simulation into account [54]:

$$\begin{aligned} B_{Hurkx} &= \sqrt{\pi}\frac{F}{F_\Gamma}exp\left[\frac{F}{\sqrt{3}F_\Gamma}\right]^2\left(2 - erfc\left[\frac{1}{2}\left(\frac{F_\Gamma E_n}{FkT} - \frac{F}{F_\Gamma}\right)\right]\right) \; for \; \frac{F}{F_\Gamma} \leq \sqrt{\frac{E_n}{kT}} \\ B_{Hurkx} &= \sqrt{\pi\frac{F}{F_\Gamma}\frac{E_n}{kT}}exp\left[-\frac{E_n}{kT} + \frac{F}{F_\Gamma}\sqrt{\frac{E_n}{kT}} + \frac{F}{F_\Gamma}\sqrt{\frac{E_n^3}{(kT)^3}}\right] \cdot \\ & \quad erfc\left[\left(\frac{E_n}{kT}\right)^{1/4}\sqrt{\frac{F}{F_\Gamma}} - \left(\frac{E_n}{kT}\right)^{3/4}\sqrt{\frac{F_\Gamma}{F}}\right] \; for \; \frac{F}{F_\Gamma} > \sqrt{\frac{E_n}{kT}} \\ F_\Gamma &= \frac{\sqrt{8m^* \cdot (kT)^3}}{eh} \\ E_n &= 0 \; for \; kTln(n/n_i) > 0.5E_g \\ E_n &= 0.5E_g - kTln(n/n_i) \; for \; E_t \leq kTln(n/n_i) \leq 0.5E_g \\ E_n &= 0.5E_g - kTE_t \; for \; E_t > kTln(n/n_i) \end{aligned} \tag{C.3}$$

The term m^* in F_Γ is an approximation of the effective mass. It is the product of a fitting factor, depending on the used approximation, and the electron mass.

C.6 Perimeter Leakage Currents

Gate induced drain currents for high-k transistors with thick gate dielectric, and the best leakage (etch samples set: 3keV Ge) are shown (Fig. C.14).

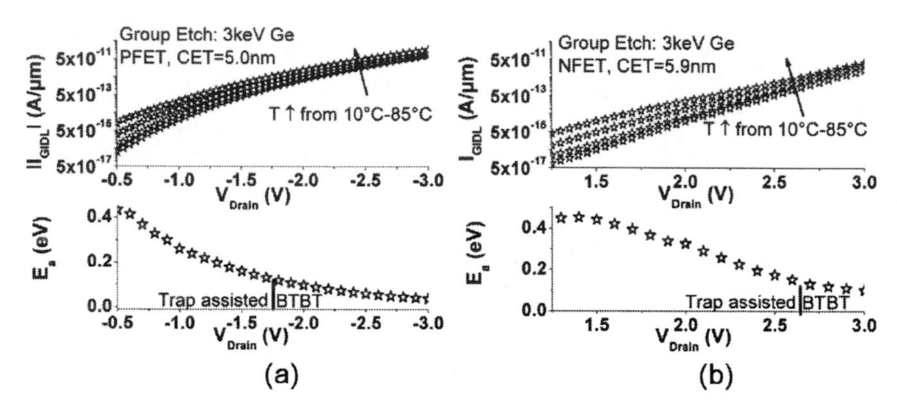

Figure C.14: The measured temperature dependence of the GIDL current for thick gate dielectric transistors from the etch set is shown in the upper part of the figure. The activation energy is plotted in the lower part of the figure. The main leakage mechanisms is changing from a trap assisted tunneling to a band to band tunneling when the bias increases.
(a) PFET
(b) NFET

C.7 Comparision of High-k and SiON Reference Devices

The leakage current of the SiON reference transistors is compared to the leakage of the high-k devices in this section.

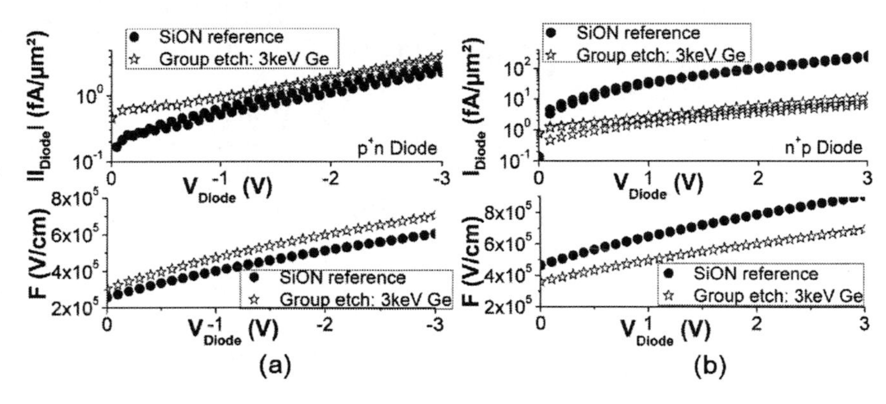

Figure C.15: Measured diode leakage current at 85°C for SiON reference devices is shown, and compared to the high-k junction leakage using a diodes of the etch sample set with a 3keV Ge implant. Five dies are measured for each sample.
(a) p^+n diode
(b) n^+p diode

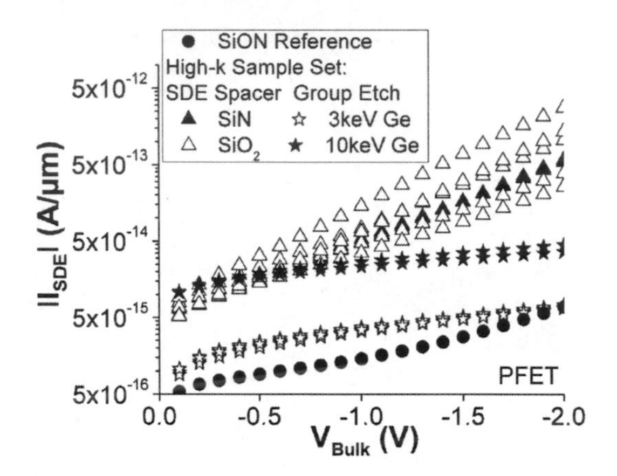

Figure C.16: SDE leakage comparison for the high-k-, and SiON reference PFETs with thin gate dielectric at 85°C. The SiON SDE leakage below 1V exhibits a lower increase with the voltage, and is possibly governed by current generated at the N_{itedge}. The SiON reference current at high voltages is probably governed by leakage generated at N_I. Five dies are measured for each sample.

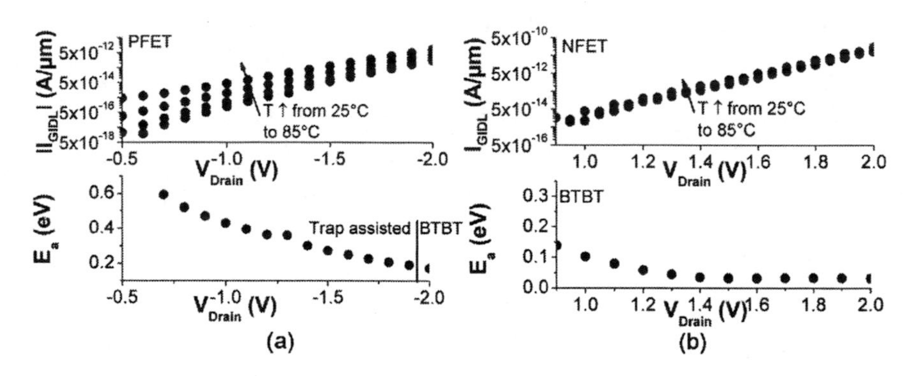

Figure C.17: Temperature dependent GIDL current (upper part of the figure), and activation energy (lower part of the figure) of the SiON reference transistors with a CET of about 2.5nm is shown for a typical:
(a) PFET.
(b) NFET.

C.8 Gate Leakage Current

Analysis of the gate leakage currents for the device from the etch sample set with a 10keV germanium implant are shown. The gate to channel-, gate to bulk-, and gate to source/drain overlap leakage is separated (Fig. C.18, C.19). Also the temperature dependence of the gate leakage is investigated (Fig. C.20).

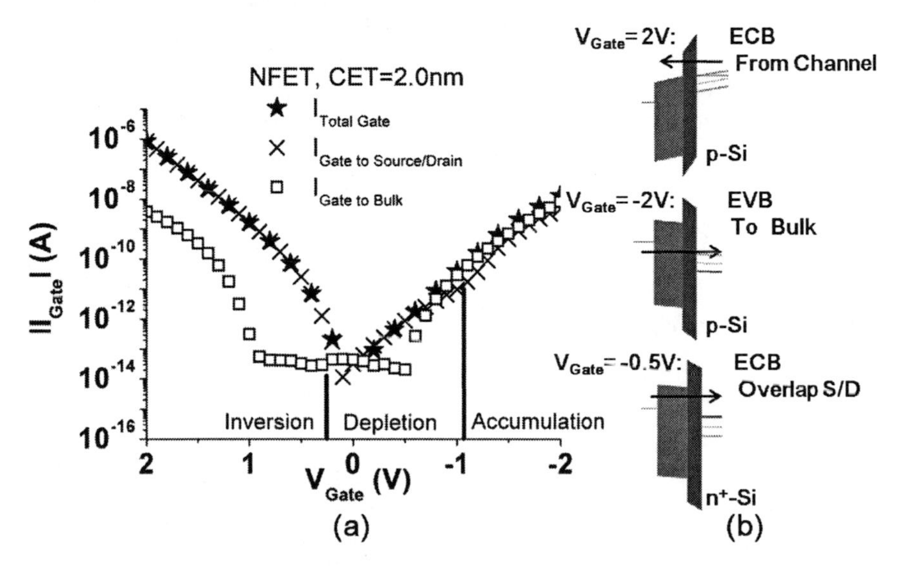

Figure C.18: The gate leakage of a 0.8μm long NFET from the etch sample set (10keV germanium) at 85°C is presented.
(a) Contribution of gate to source/drain, and gate to bulk leakage on overall gate current depending on the voltage are shown for the thin gate dielectric device.
(b) Sketch of the leakage flow due to electron conduction band tunneling (ECB) at V_{Gate}= 2V, and electron valance band tunneling (EVB) at V_{Gate}=-2V is presented. The source/drain overlap current dominates the depletion regime (V_{Gate}= 0.5V) [35].

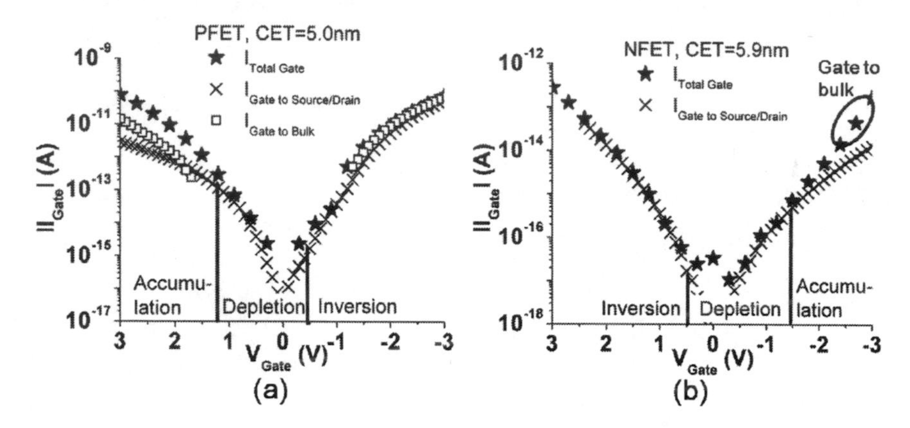

Figure C.19: The gate leakage of a 0.8μm long devices from the etch sample set (10keV germanium) at 85°C is presented. The contribution of gate to source/drain, and gate to bulk leakage on overall gate current depending on the voltage are shown for the thick gate dielectric devices.
(a) PFET
(b) NFET

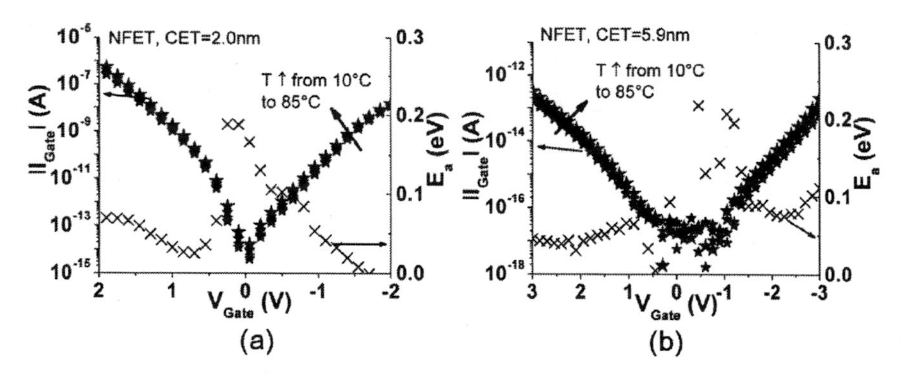

Figure C.20: The temperature dependence of the gate leakage for 0.8μm long NFETs from the etch sample set (10keV germanium) is presented for:
(a) A device with thin gate dielectric.
(b) A device with thick gate dielectric.

C.9 Simulation Parameter of the BSIM Model Card

Table C.10: BSIM Parameters of the Gate Leakage

Parameter	PFET SiO_2 Spacer 10keV Ge	PFET SiN Spacer 3keV Ge	NFET SiN Spacer 3keV Ge	Gate Leakage
EOT (nm): L_{Gate}=100nm	2.08	2.11	3.35	all
EOT (nm): L_{Gate}=10μm	1.97	2.11	2.58	
AIGBACC ($F^{0.5}sg^{-0.5}m^{-1}$)	0.019	0.018	0.0137	Gate to Bulk Accumulation
BIGBACC ($F^{0.5}sg^{-0.5}m^{-1}V^{-1}$)	0.0040	0.0035	0.002	
CIGBACC (V^{-1})	0.002	0.002	0.002	
NIGBACC	1	1	1	
AIGBINV ($F^{0.5}sg^{-0.5}m^{-1}$)	0.0183	0.0158	0.0133	Gate to Bulk Inversion
BIGBINV ($F^{0.5}sg^{-0.5}m^{-1}V^{-1}$)	0.065	0.05	0.029	
CIGBINV (V^{-1})	0.002	0.002	0.002	
NIGBINV	3	3	3	
EIGBINV (V)	1.1	1.1	1.1	
AIGC ($F^{0.5}sg^{-0.5}m^{-1}$)	0.0135	0.0127	0.02	Gate to Channel Inversion
BIGC ($F^{0.5}sg^{-0.5}m^{-1}V^{-1}$)	0.022	0.019	0.035	
CIGC (V^{-1})	0.002	0.002	0.002	
NIGC	1	1	1	
AIGS/AIGD ($F^{0.5}sg^{-0.5}m^{-1}$)	0.0118	0.0099	0.015	Gate to S/D
BIGS/BIGD ($F^{0.5}sg^{-0.5}m^{-1}V^{-1}$)	0.0015	0.0015	0.003	
CIGS/CIGD (V^{-1})	0.002	0.002	0.002	
NTOX	1	1	1	
POXEDGE	1	1	1	

C.10 Fit Parameters for the Subthreshold Current

The geometry parameter (η) is expected to be constant (equation 2.5). No significant change in η for the geometrical similar transistors can be found, so the fit is done correctly.

Table C.11: Parameters of Subthreshold Current of 100 Long Transistors (CET≈2nm)

PFET		V_{th} (V)*	λ**	m*	$(\eta\epsilon_{Si})/(L_{Gate}\epsilon_{ox})$
SiON		0.57	0.04	1.5	0.011
Group Etch	3keV Ge	0.67	0.055	1.28	0.020
Group Etch	10keV Ge	0.66	0.055	1.28	0.020
Group Spacer	SiN	0.63	0.075	1.34	0.017
Group Spacer	SiO_2	0.62	0.045	1.44	0.023
NFET		V_{th} (V)*	λ**	m*	$(\eta\epsilon_{Si})/(L_{Gate}\epsilon_{ox})$
SiON		0.52	0.03	1.36	0.008
Group Etch	3keV Ge	0.45	0.065	1.38	0.024
Group Etch	10keV Ge	0.44	0.065	1.40	0.023

* Extracted from the transfer curves at $V_{Drain}=\pm0.05V$.

** Extracted from the fit of $I_{SubVth}V_{Drain}$.

C.11 Basic Electrical Characteristics of Long Channel Devices from the Sample Set with Different Spacers

Table C.12: Basic Characteristics of PFETs with Different SDE Spacers(L_{Gate}=800nm, CET≈2nm)

	SiO_2 Spacer	SiN Spacer
CET (nm)	2.08±0.02	2.04±0.02
L_{ov} (nm)	4.4±0.2	3.0±0.9
V_{th} (V) V_{Drain}=-0.05V (CV**)	-0.55±0.05	-0.55±0.05
V_{fb} (V) (CV**)	0.5±0.05	0.5±0.05

All values are given for measurements on 5 dies (wafer center). All PFETs have a 10keV Ge preamorphization prior to high-k removal.
* Capacitance voltage measurement are done a diode structure with SDE and S/D implants.
** V_{fb} and V_{th} are taken from the point of maximum, and minimum slope of the CV characteristic.

Table C.13: Basic Characteristics of PFETs with Different SDE Spacers (L_{Gate}=800nm, CET≈5nm)

	SiO_2 Spacer	SiN Spacer
CET (nm)	5.08±0.06	5.01±0.13
L_{ov} (nm)	9.7±1.1	8.3±0.2
V_{th} (V) V_{Drain}=-0.05V (CV**)	-0.6±0.05	-0.5±0.05
V_{fb} (V) (CV**)	0.45±0.05	0.55±0.05

All values are given for measurements on 5 dies (wafer center). All PFETs have a 10keV Ge preamorphization prior to high-k removal.
* Capacitance voltage measurement are done a diode structure with SDE and S/D implants.
** V_{fb} and V_{th} are taken from the point of maximum, and minimum slope of the CV characteristic.

C.12 Gate Leakage of the Thick Gate Dielectric PFETs

The unusual high gate leakage for the thick gate dielectric PFETs with silicon nitride spacer is reduced by improving the etching processes between the two sample sets.

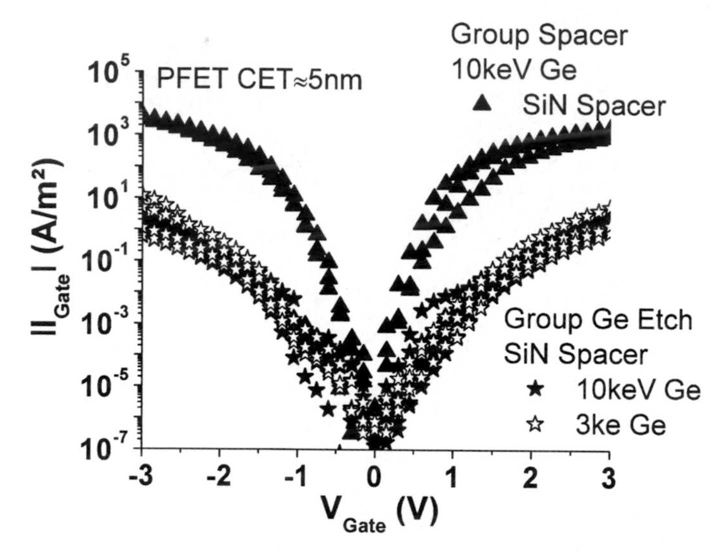

Figure C.21: Gate Leakage current for PFETs with thick gate dielectric depending on the bias at 85°C. Measurements of five dies each for the samples sets with different spacers, and with different germanium etching implants are shown. The gate leakage is improved by changes in the etching process between the two sample sets.

C.13 Basic Electrical Characteristics of Long Channel Devices from the Sample Set with Different High-k Etches

Table C.14: Basic Characteristics of Transistors with Different High-k Etches (L_{Gate}=800nm, CET≈2nm)

	PFET		NFET	
	10keV Ge	3keV Ge	10keV Ge	3keV Ge
CET (nm)	2.18±0.01	2.19±0.02	2.22±0.04	2.22±0.02
L_{ov} (nm)	3.5±0.3	3.9±0.5	4.9±0.2	4.9±0.1
V_{th} (V) V_{Drain}=-0.05V (CV**)	-0.55±0.05	-0.55±0.05	0.38±0.09	0.40±0.05
V_{fb} (V) (CV**)	0.5±0.05	0.5±0.05	-0.72±0.09	-0.73±0.08

All values are given for measurements on 5 dies (wafer center). All devices have a SiN SDE spacer.
* Capacitance voltage measurement are done a diode structure with SDE and S/D implants.
** V_{fb} and V_{th} are taken from the point of maximum, and minimum slope of the CV characteristic.

Table C.15: Basic Characteristics of Transistors With Different High-k Etches (L_{Gate}=800nm, CET≈5nm)

	PFET		NFET	
	10keV Ge	3keV Ge	10keV Ge	3keV Ge
CET (nm)	5.11±0.03	5.10±0.04	5.91±0.01	5.89±0.04
L_{ov} (nm)	8.1±0.3	8.7±0.7	13.3±0.2	12.4±0.3
V_{th} (V) V_{Drain}=-0.05V (CV**)	-0.55±0.05	-0.55±0.05	0.46±0.07	0.45±0.05
V_{fb} (V) (CV**)	0.5±0.05	0.5±0.05	-0.60±0.05	-0.63±0.08

All values are given for measurements on 5 dies (wafer center). All devices have a SiN SDE spacer.
* Capacitance voltage measurement are done a diode structure with SDE and S/D implants.
** V_{fb} and V_{th} are taken from the point of maximum, and minimum slope of the CV characteristic.